ARRL's

Low Power Communication

The Art and Science of QRP

Richard Arland, K7SZ

Published by:

ARRL *The national association for AMATEUR RADIO*

225 Main Street • Newington, CT 06111-1494 USA

***ARRLWeb*: www.arrl.org/**

Printed in USA

ISBN: 0-87259-917-5

Second Edition
First Printing, 2004

Contents

Dedication

Mike Branca, W3IRZ

Born: March 6, 1941 *Died: September 29, 2003*

On September 29, 2003, a rift appeared in the fabric of QRP. On that day, we lost a very dear member of our unique and close-knit fraternity. Mike Branca, W3IRZ, passed from this Earth after an extended illness.

When I first met Mike, I knew that I was in the presence of a True QRPer. His demeanor and his humor shone like a beacon in the night. During a family trip to Atlanta, a group of the North Georgia QRPers had arranged an afternoon get together at a local restaurant. Mike was one of those present and, after a brief round of introductions, we all started talking QRP. Without fanfare, Mike produced one of his many homebrew projects and handed it to me. I was holding a completely homebrew version of the NorCal Sierra multi-band transceiver. The workmanship was incredible. His attention to detail was amazing. During the ensuing conversation, Mike told me of several circuit modifications he had made to the original design to improve the operating characteristics of the rig. I was blown away.

Over the next several years I managed to meet and talk with Mike a number of times when my travels took me to the Atlanta area. The last time I met him he handed me a small white box and told me to have fun building the kit. This project was one of his prolific designs; a two-tube, 80-meter transmitter called "The Twin-Tube Eighty." Mike loved to build and tinker with all things QRP. He greatly enjoyed sharing his passion with others and was not above handing someone a kit to get them started in the right direction. The low-power side of the hobby was his first love and building and operating homebrew gear made it all that much more enjoyable. This is what he loved to convey to others; a passion for building and the thrill of operating QRP.

Mike's dedication to the ham radio hobby in general and QRP in particular is exemplary. He personified my idea of what a true QRPer is all about. His involvement with the North Georgia QRPers helped the group grow and expand into one of the best regional clubs around. Never tiring of sharing his technical prowess and always quick with a joke, Mike embodied the spirit of QRP.

Mike, I miss you. We all do. My life, along with many others, has been enriched by your presence. Rest in peace, my friend.

Foreword

In the past, QRP operation was restricted to CW using a homebrew two-transistor transmitter and a receiver as wide as 40 meters itself. Today, QRP is hot! It's one of the most popular activities on the air today. Gone are the days of the homebrew drift-o-matic transmitter. Now we use DDS generated signals, DSP processed audio and the latest in computer control.

Homebrewing is still a mainstay of QRP operation, but now we are building with surface-mount parts, programmable PIC controllers and hybrid amplifiers on a stick.

While it's true that CW provides the most bang for the watt, we have pushed the envelope of communications with such modes as packet, PSK31 and even satellites. Of course, good old SSB is now a favorite with some QRP operators.

To me, one of the best parts of running QRP is portable operation. Just you, the rig, some wire and a solar panel. For some really great fun, drop your power down to five watts and join the HFpack group on 18.1575 MHz USB. Find out for yourself just how far you can talk with QRP.

To make sense of all this, Rich Arland, K7SZ, has once again gathered all the information you need to know into one easy-to-read book. So no matter if you want to pick out a vintage QRP transceiver on eBay, or need tips on how to work contests running QRP, Rich has the answers. This book is your guide to getting the most out of your QRP signal. Have fun. I'll look forward to hearing you on the bands!

Mike Bryce, WB8VGE

Acknowledgements

This is the place where I get to say "Thank You" to the various people that helped make this book a reality.

Once again, my fantastic wife, Patricia, who still hasn't gotten all of her drywall up and spackling done! Honest, honey, I *will* get the dining room done! So far she's made it through five books without a whimper. You just can't ask for a better life-mate than that!

To Mike Bryce, WB8VGE, Ed Breneiser, WA3WSJ, Marshall Emm, N1FN, Bob Locher, W9KNI, Ed Wetherhold, W3NQN, Dave Benson, K1SWL, George Heron, N2APB, Joe Everhart, N2CX, Cam Hartford, N6GA, Fran Slavinski, K3BX, Paul Stroud, AA4XX, Bob Witte, KØNR, Andy Howard, WA4KCY, Dar Piatt, W9HZC, Tim Cook, NZ8J, Bob Hightower, KI7MN, Roy Lewallen, W7EL, Tom Dandrea, N3EQF, the staff at MFJ Enterprises, Ten-Tec, Yaesu (Vertex Standard), ICOM, Kenwood, Elecraft, SGC, Inc., Wilderness Radio, Oak Hills Research and Small Wonder Labs for the help and background information needed to do this book.

To all you QRPers and soon-to-be-QRPers everywhere. Thank you all for supporting my first book and my "QRP Power" column in *QST*. Your efforts and dedication to the craft inspire us all.

Last but not least, to Larry Wolfgang, WR1B, my editor on the two books I've written for ARRL. He has the patience of a saint and manages to pull all my mundane ramblings together to make a tremendous manuscript come alive. Larry, you're the best!

Richard Arland, K7SZ

About the Author

Rich Arland, K7SZ (ex K7YHA), of Wilkes Barre, Pennsylvania, has been a ham and avid QRP enthusiast since 1963. He currently holds an Amateur Extra ticket. In addition to his passion for anything to do with QRP, Rich also enjoys restoring and using older tube-type equipment (boat anchors).

Rich is a retired Air Force Master Sergeant with 20 years in military communications, including long-haul and tactical units. While in the service, he got to be DX during his three years in the Azores (CT2BH) assigned to the 1936 Communications Squadron. Rich also spent time in England (G5CSU), Japan (KA2AA) and Germany (DA2NE) during his military career.

(WA4KCY photo.)

He has worked as a broadcast engineer, and currently teaches basic and advanced vocational electronics at the state correctional institution in Dallas, Pennsylvania. He also serves on the state board that oversees video surveillance upgrades for all Pennsylvania prisons. He is the author of numerous *QST* and ARRL Web Extra articles. Rich served as editor of the *QST* QRP Power column for three years. Rich has been an ARRL Official Emergency Station since 1991, and is also an ARRL Technical Advisor.

Prolog

In 1985 Adrian "Ade" Weiss, W0RSP (ex K8EEG) wrote the very first book dedicated entirely to low power communications, ***The Joy of QRP***. It became an instant classic. Although by today's standards it is quite dated, almost 20 years later QRPers still buy and read this informative book. Having been an active QRPer since 1965, I found Ade's book a tremendous help and I gained much valuable information from reading and re-reading it. ***The Joy*** became my reference and I vowed that should I ever have the chance, I would contribute what I could to further the QRP hobby by writing articles and possibly a book on the subject. Kind of a repayment for what Ade had done for me and countless other QRPers in the past.

When I wrote the first edition of ***The ARRL's Low Power Communications, The Art and Science of QRP*** in late 1997, I was not really aware of how much this book was needed in the Ham Radio community. The book represented the first fresh, new, up-to-date treatment of low power communications in well over ten years. Sales figures the first year were great, indicating that ***Low Power Communications*** (***LPC***) was destined to be an ARRL top seller. Sales records over the last several years bear witness to the fact that radio amateurs are up to the challenges that QRP offers, making QRP the fastest growing facet of the Ham Radio hobby.

A lot has changed in the QRP hobby over the last five years. New gear, interesting antenna designs, new modes, new homebrew construction methods, new contests and operating events, all pointed to the need for an updated and completely revised edition of ***LPC***.

The *only* guarantee in life is that life is guaranteed to change! And so it is with QRP. The low power side of the hobby has undergone some tremendous changes, which broadens its allure to the Ham Radio hobbyist. No longer can QRPers (those who practice the craft of low power communications) be relegated to the fringe element of the hobby. Instead, QRPers are on the cutting edge of technology, leading the way for the rest of the Radio Amateur Fraternity. That's a pretty weighty statement. And, yes, it's true!

For instance, QRPers Wayne Burdick, N6KR, and Eric Swartz, WA6HHQ, combined their talents and formed a company called Elecraft. Their major contribution to the QRP hobby is a little transceiver kit called the **K2**. This kit radio has revolutionized homebrewing and ham radio. Costing less than $600 (for the basic CW kit), a homebrewer with intermediate building skills can construct an all-band CW rig that will compete with the top-of-the-line imported radios. What? You don't believe me? Check out the March 2000 *QST* product review of the Elecraft K2. Wayne and Eric also provided a whole stable full of options that greatly expand the capabilities of the K2, resulting in a CW/SSB/Data radio with built in auto antenna tuner, internal battery, computer interface, DSP filter and an external VHF transverter. QRPers aren't the only people flocking to the K2 camp. High power operators, especially contesters and VHF/UHF weak signal operators are so impressed with the receiver performance of the K2 that they are using this inexpensive super-performer to drive RF linear amplifiers and as IF strips for VHF/UHF transverters. With the addition of the KPA100 100-watt amplifier kit, the K2 can now be two radios in one: a high-performance QRP radio and a full-featured 100-W rig.

Not to be left resting on their laurels, Team Elecraft produced a scaled down version of the K2, appropriately called the **K1**, in 1999. This is a CW only, multi-band transceiver kit that includes many of the features found on the K2. The price of the dual-band K1 is only $279. Options include internal auto antenna tuner, internal battery, noise blanker, and special tilt-stand for portable use. The K1 is geared for the on-the-go QRPer who needs a small, compact, rig for camping, hiking, business trips or family outings.

The K1 and K2 have made a definite impression on the Amateur Radio community. So much so that recently Yaesu, Icom and Ten-Tec have all released their own QRP radios. This may not seem to be such a big deal, but in talking with the folks at Ten-Tec and Yaesu, this is a major step for both companies. It costs a bundle of money for a well established, mainstream commercial company to retool and set up to manufacture a new radio. This is on top of the research and development costs of the new rig. It's also a big risk, since entering a new market (QRP) can be a little scary. Until recently, QRP did not figure in their product lines. Now it does. This points once again to the fact that QRP has emerged into the mainstream of the ham radio hobby, and QRPers are a fresh market for new gear and accessories.

At the opposite end of the spectrum is Dave Benson, K1SWL, of Small Wonder Labs in Connecticut. In the mid 1990s, Dave introduced a series of inexpensive minimalist CW transceiver kits, followed by a series of PSK31 transceivers and finally the **RockMite** CW transceiver kit that takes up less room than a pack of cigarettes. Costing only $25 for the kit (not including a case, connectors and knobs), the **RockMite** has become a cult radio with a tremendous following. The average QRPer ten years ago could not build a kit like the **RockMite**, simply because the technology didn't exist at the time.

For some inexplicable reason, we are intrigued by the technology of our hobby and we all love the hardware aspect of radio. There is another side of the hobby to be investigated, however: operating modes. Traditionally, QRPers have relied on CW, using the International Morse Code, to communicate almost effortlessly, on a worldwide scale. The QRP fraternity did not look upon data and voice modes with favor until recently. With the advent of digital signal processing, improved audio circuitry in the radios and the invention of new modes like PSK31, QRPers now have several alternatives to See-Double-You!

Since the restructuring of the Amateur Licensing criteria in April of 2000, there are thousands of newcomers on the HF bands. Many of these folks had Novice or Technician licenses but had not upgraded because they did not like CW or were uncomfortable using it. Now, with the new license criteria firmly in place, we have a whole new group of potential QRP operators just begging for information, direction and help. These neophyte HF operators are in search of fun, excitement and a challenge. Let's grab them now and get them started in QRP. In this newest edition of ***LPC,*** I will cover in detail the use of voice (SSB) and data (PSK31) as it applies to QRP operating. This should definitely appeal to these new HF operators. So, you see, within the QRP hobby there is something for everyone.

Building your own station gear (homebrewing) and QRP are inseparable. They are inexorably joined at the hip like a set of Siamese twins. Seldom will you find a QRPer who has not built some or all of his or her station. Whether it's only a simple wattmeter or an entire transceiver with accessories, QRPers are homebrewers, plain and simple.

Over the last few years there has been a tremendous upsurge in homebrewing within the QRP community. The remarkable transceiver kits that started appearing in 1994 initiated this. The NorCal-40 and 40A, the Sierra (by Wilderness Radio), the Small Wonder Labs SW-40 and Green Mountain kits, Oak Hills Research's OHR 100 and 500 kit radios and finally Elecraft, all contributed to the rapid rise of homebrewing within the ham radio hobby. While American hams think this is a recent development, in Europe, especially the United Kingdom, homebrewing has been going on for many, many years. We Americans are starting to catch up with our friends across the pond, however.

One building technique that has gained rapid acceptance is called Manhattan Construction, named for the small pads glued to the surface of the printed circuit board material used as a chassis. If you look at the chassis at the proper angle (and apply a liberal dose of imagination), the PC layout starts looking like the Manhattan skyline, complete with skyscrapers and streets! This method of "Ugly Construction" (a term coined by Doug DeMaw, W1FB — now a silent key) yields a neat-looking project, which, because of the continuous ground plane furnished by the PC board material, is almost guaranteed to work the first time, every time. This kind of success rate is a great confidence booster to beginning and intermediate homebrewers wanting to gain experience and further their electronic skills. You can literally build almost any type of circuit using Manhattan Construction. We'll look at Manhattan Construction with a couple of simple projects guaranteed to whet your homebrewing appetite.

One of my pet peeves is a lack of coverage of the operating side of the low power hobby. It seems that there is a plethora of circuits, projects and kits available to the QRPer but little information on how to communicate using less than 5 watts of

RF. This book deals with that void. In the first edition, I purposely chose to exclude any QRP hardware projects and concentrate on the "how and why" of QRP, in order to bring some balance to the information available to the newcomer in low power communications. After all, you can build and build all sorts of transceivers and accessories, but you aren't communicating with anyone. (No, the Internet doesn't count!) It's not until you fire up one of your creations and finally get on the air with it, that you become a true QRPer.

I hope you enjoy this book. I wrote it with much love of the hobby in mind. Thanks to all of you who purchased the first edition and those of you who were loyal readers of my **QRP Power** column each month in *QST*. Thanks, also, to all of you who have provided comments, complaints, ideas and shared your thoughts with me regarding QRP. I dedicate this book to you. Finally, "Thanks Ade", for all your help, guidance and encouragement during the lean years.

CHAPTER 1

Introduction to Low Power Operating

In this chapter we are going to take some of the most Frequently Asked Questions (***FAQs***) about QRP and provide some answers for the low power communicator.

FAQ: WHAT DOES "QRP" MEAN?

When we talk about ***QRP*** what are we *really* talking about? Most radio amateurs recognize ***QRP*** as one of the many "Q" signals used in radio telegraphy as an abbreviation. Literally translated, ***QRP*** means: "Please reduce power," or when coupled with an interrogatory ("?"): "Shall I reduce power?"

FAQ: WHEN WAS THE TERM "QRP" COINED TO MEAN UNDER-FIVE-WATT HAM RADIO?

It is difficult to pinpoint when the term ***QRP*** was first used to denote low power (less than five watts output) communications. Since this term was in vogue when I first became active in the low power arena in 1965, I think it is safe to assume that the use of the term to denote low power communications has been around for well over 40 years. However, today ***QRP*** is used to refer to communications using five watts or less RF power output and ***QRPers*** are those individuals who practice the craft of under-five-watt communications. For those adventurous few who want to really push the envelope and engage in *milliwatting*, the term ***QRPp*** is used. So, there you have it: ***QRP*** is 5 W or less and less than 1 W is identified as ***QRPp***.

***Sherry Dease, KC4WYL, operated from this artistic setting on Keanae Peninsula, Maui, Hawaii during the 1998 Field Day.* (Photo courtesy of N4NR.)**

FAQ: IS QRP A NEW MODE OR HAS IT BEEN AROUND A WHILE?

The history of ***QRP*** dates back to the earliest days of radio. Many radio amateurs got on the air with minimal rigs built from junk box parts and old Model "T" Ford ignition coils. Receivers of that period were grossly inefficient, requiring the radio experimenter to keep increasing his spark output in order to communicate over greater distances. Interference on the bands (at that time 200 meters and longer wavelengths) was horrendous. If you have ever heard the raspy, almost evil snarl of a spark gap transmitter you'll get the picture!

When spark gave way to CW, the vacuum tubes of the day were not the high power tubes of the late 1940s, 50s and 60s. The Armstrong regenerative detector, and later his superhetrodyne receiver designs, greatly improved the sensitivity of the receiving equipment in the early 1920s. Transmitting circuitry, on the other hand, was not optimized and efficiency of the early CW transmitters was very low, resulting in greatly reduced output even though the input power to the final amplifier might be given in three or four digits. Many of these early radio experimenters practiced the craft of low power communications by default!

As these fledgling radio amateurs continued their experiments with CW it became very apparent that really high power was not needed on the bands with wavelengths shorter than 200 meters. Little was understood about HF propagation at that time, so there was a lot of trial and error involved, but the one message that kept coming through was that the "Ether Busters" and "Watt Hogs" had no place in modern ham radio. They were replaced by "The New

American Amateur," a radio experimenter who valued equipment and antenna efficiency along with superior operating skills over raw RF power.

QRP GETS A BOOST FROM UNCLE SAM!

In the 1940s, the US Military started using a QRP rig in the form of the BC-611 "handie-talkie." This quart-milk-carton sized HF transceiver (most were on 80 meters, around 3885 kHz) put out a massive 300 milliwatts of RF energy into a very short whip antenna. Made initially by Motorola, these first tactical hand-held transceivers provided desperately needed front-line communications at the squad/platoon level.

My good friend, Andy Howard, WA4KCY, has seven of these little beauties along with a poster showing a BC-611, all muddy and grimy, sitting next to a brand new Motorola "flip-fone" cellular telephone. The caption reads: "What did you do in the war, Grandpa?" It is safe to say that radio communications (including RADAR) had the same tremendous impact on WWII that the airplane had on WWI, and a QRP rig was at the center of the action!

After WWII, the abundance of military surplus radio gear along with really high power transmitting tubes provided the frugal radio amateur access to the high power world of ham radio. QRP was overshadowed by high power radio gear for some time.

THE MODERN QRP MOVEMENT

In 1961 the congestion on the HF bands prompted Harry Bloomquist, K6JSS, to found the QRP Amateur Radio Club International (QRP ARCI), in an attempt to bring some sanity to the escalating power cravings of many hams. While the QRP ARCI was founded ostensibly to encourage members to voluntarily limit their power ***input*** to 100 W or less, this was not a "true QRP club" as defined today. It took about 18 years and a lot of effort by forward thinking members to transform the QRP ARCI into a real QRP club, recognizing the RF power *output* limit of only five watts.

During this time the transistor was making inroads into ham radio gear construction. Occasionally, *QST, CQ* and *Ham Radio Magazine* would offer a simple transistorized QRP transmitter circuit for the homebrewer. International Crystal in Oklahoma

City offered simple one-transistor oscillators, mixers and associated small printed-circuit boards to assemble low power gear. There were a lot of flea-powered transistor rigs on the air during this time, thanks to International Crystal.

Then, in the late 1960s, Ten-Tec in Sevierville, TN, revolutionized the QRP hobby by manufacturing and marketing the first commercial QRP transceiver, the PM-1. This rig also came as a kit with four PC boards that you could mount in your own chassis. Since that time, Ten-Tec has been a close friend of the QRP community by marketing many first class QRP transceivers and accessories.

Ten-Tec followed their PM-1 in rapid succession with the PM-2, and PM-3 series transceivers. There were some slight changes in band coverage and the PM-3a offered a semi-break-in keying circuit. The Sevierville group also offered several accessories: an ac power supply, SWR bridge and antenna tuner to match the PM-series.

Not to be outdone, in 1972, Heathkit produced the first in a series of three "Hot Water" QRP rigs, the HW-7. Costing only $79.99 this three band (40/20/15 meters) kit gave many of us QRP Old Timers our first commercially designed QRP kit radio. The HW-7 had a *horrendous* direct conversion (DC) receiver and was never really up to the task of operating on 40 meters or anywhere near an AM broadcast station, but they were a fun radio, nonetheless.

In 1976 the Heathkit design team came out with the HW-8, which was continuously in production for the next seven years, until 1983. The "Hot Water Eight" was a very popular QRP rig and they are much sought after by collectors today. At a cost of $139.95 new, the HW-8 provided the frugal QRPer with four bands (80 - 15 meters), full VFO control, a very stable and much improved direct conversion (DC) receiver *and* an internal active audio-frequency (AF) filter. While neither the HW-7 or 8 was capable of anything close to full-break-in (QSK) keying, these radios were a great way for newcomers to the hobby to start enjoying QRP without plunking down a ton of money.

***Mark Calderazzo, WB4UOK, operates a simple station during Field Day.* (Photo courtesy of WB4UOK.)**

In 1984 Heath released their final QRP rig, the HW-9. Not nearly as popular as the HW-8, the 9 offered nine band HF coverage with the purchase of an optional band pack. Break-in keying was much improved but still was not full QSK. Also much improved was the receiver. Heath dropped their DC receiver in the previous models in favor of a superhet design. Although plagued with various problems, the HW-9 was a good solid QRP CW transceiver; once some modifications were done, this rig played pretty well. Costing $249.95 without the optional band pack, these rigs have come to the attention of collectors.

AND FOR ALL YOU SSB OPERATORS OUT THERE...

Starting in the mid 1970s Ten-Tec released three analog Argonaut SSB/CW transceivers in succession: the model 505, 509 and 515. All three of these rigs covered 80/40/20/15 and 10 meters and offered, what has become the industry standard for flawless, utterly beautiful full break-in keying. Once you use any of the Argos, and experience their ultra smooth QSK, you'll be hard pressed to go back to anything else.

THE OFF-SHORE EMPIRE STRIKES BACK!

By the mid-80s both Yaesu and Kenwood imported their low power (10-W) versions of their very popular low-end SSB/CW transceivers. The Yaesu FT-301S was the only rig of that period that would cover 160 meters, for those who enjoyed "Top Band." The Kenwood TS-120S and TS-130S radios became very popular with QRPers. It was simple to crank the RF drive back and enjoy operating QRP (both CW and phone) using a very well designed commercial transceiver. These two sets still bring top dollar in the collecting game.

THE CURRENT QRP EXPLOSION

All this set the stage for the explosion of QRP operation and homebrewing that erupted in the early to mid 1990s. The QRP ARCI became an umbrella organization as regional QRP clubs sprang up all over the states. Across the pond in Europe, the United Kingdom G-QRP Club has dominated the QRP hobby from the late 1970s to present. There are organized QRP groups all over the world. Many sponsor contests, sprints and operating periods on a regular basis.

Under-5-watt ham radio has undergone tremendous changes over the last 40 plus years and has emerged as a mainstream portion of the ham radio hobby. Why? Because QRP is challenging, intriguing, very demanding and a lot of ***F-U-N*!**

WATTS VERSUS S-UNITS

FAQ: HOW CAN YOU COMMUNICATE USING ONLY FIVE WATTS?

When we speak of received signal-strength level (RSL) we immediately default to the S-meter on the front of the radio. Why? Because its there! The idea of using a calibrated meter to measure incoming signal strength had its merits. Unfortunately, over the years, the relationship between the S-Unit and the associated microvolt signal input and power ratio (in decibels or dB) has become blurred. It's time to redefine exactly what an S-Unit really is and how we can gain some meaningful information from what we see on the S-meter.

Originally S-Units were calibrated in microvolts:

S-1	0.2 µV	S-6	6.0 µV
S-2	0.4 µV	S-7	13.0 µV
S-3	0.8 µV	S-8	25.0 µV
S-4	1.6 µV	S-9	50.0 µV
S-5	3.0 µV		

***Mike Gaude, WK6O, hangs a simple wire antenna to operate during a backpack camping trip.* (Photo courtesy of WK6O.)**

As you can see, there is a distinct relationship between the "S" value and the associated microvolt signal applied to the receiver's antenna input terminals. Each S-Unit increase (or decrease) results in a two fold increase (or decrease) in the microvolt signal level. This linear scale is handy for engineers but not for mainstream amateur radio operators who are used to dealing in watts. What *we* need is a scale referencing the S-Unit to actual RF power increases or decreases. Such a scale has the added benefit of showing how well QRP will work

Luckily we have one! Since decibels (dB) are a logarithmic function, each S-Unit equals a 6-dB power increase (or decrease). What this actually means is that if you want to show a 1 S-Unit increase in signal strength, you'll need to increase your power by four times or 6 dB. Therefore, if you have a 10-W transmitter and the RSL at the distant end station is an S-5, you'll have to increase your power output to 40 W just to gain one S-Unit (S-6)! Wow, now we have something to play with!

OK, so now you know the secret to S-Units. Let's take a look at a typical 100-W ham transmitter and what happens to the RSL when we start *decreasing* our RF output power to reach the Land of QRP! For arbitrary reference, let's say that when we key our 100-W transmitter up, the distant-end receiving station sees an S-9 signal on his meter. OK, that's our benchmark.

S-9	100 W
S-8	25 W
S-7	6.25 W
S-6	1.56 W
S-5	0.39 W

Whoa, Dude! Check that out! With S-9 representing a standard ham radio transmitter power output of 100 W, dropping your output power 95 W, to 5 W (QRP power levels) results in going down to slightly below an S-7 signal. That is a respectable signal on all but the most demanding band conditions and should be easily copied by anyone with basic ham radio operating skills. In reality, most of the time these same 100-W signals will be well above S-9 and your QRP signals will have correspondingly higher readings on the S-meter. For those of you who require a mathematical solution, this 95-W decrease in output power equates to 13 dB (dB= 10 [log P_1/P_2]) or slightly more than two S-Units.

FAQ: *OK, THE FIGURES LOOK GOOD, BUT DOES IT* REALLY *WORK?*

What's the moral of the story? Dropping your power down from the standard 100-W level to the QRP level of 5 W only decreases the RSL at the distant end by around two S-Units. QRP is definitely "doable"; we've just proven that mathematically. In actual practice, reliable long-haul communications using only 5 W of RF output power works quite well. Seldom will you encounter any "rock crushing" signals, but what you do hear is readily workable.

FAQ: *I READ SOMEWHERE THAT S-METERS ARE NOTORIOUSLY INACCURATE. IS THIS TRUE?*

Now for the bad news: the majority of the S-meters in today's equipment are *NOT* calibrated in accordance with the microvolt chart shown previously. The S-meter in most modern equipment is in the automatic gain control (AGC) circuit, so it measures the AGC output voltage. What we are *really* dealing with, when it comes to S-meters on our gear, is a *relative* signal strength, *not* an accurate representation of the microvolt signal appearing at the antenna input. It has also been debated that the weighting of the metering on many S-meters is not in keeping with the 6dB/Unit scale that is the industry standard.

At any rate, accurately calibrated or not, our S-meter gives us some indication of what is going on with the RSL. The main purpose in showing the relationship between RF output power and S-Units was to put your mind at ease and show that by decreasing your output by 13 dB you can still radiate a readable signal on the bands.

FAQ: *I'VE ALWAYS HEARD THAT "THE OTHER GUY" DOES ALL THE WORK IN A QRP CONTACT. IS THIS TRUE?*

Hardly! (Refer to the above.) As we've shown previously, 90% of the time we encounter a QRP signal on the air it will be very readable. Sure, there will be those times that you'll have to really work to pull the signal out of the noise due to interference (QRM), unstable or crowded band conditions, or possibly atmospheric noise (QRN), but the majority of the time QRP signals are surprisingly easy to copy. Today's receivers are light years more advanced than receivers of just ten years ago, featuring digital signal processing (DSP), razor sharp IF filtering, adaptive audio filtering and outstanding dynamic range and inter-modulation distortion (IMD) characteristics. All this enables the operator to copy signals much better. Personally, I feel that if you are unable to copy an S-5 to 7 signal under normal band conditions then you need to find another hobby!

OPERATING EVENTS

FAQ: *ONCE I GET ON THE AIR USING QRP LEVELS, WHAT CAN I EXPECT TO DO?*

Obviously, if you are serious about trying QRP, you want to

have some fun. That's where the various operating events come into play. QRP operating events have mushroomed in recent years and they come in all shapes and sizes. There is literally something for everyone. There is the ever present "rag chew": just getting on the air near one of the QRP "watering holes" and striking up a conversation with a fellow QRPer. You'll find that you will make lots of friends and have a ball just rag chewing.

If you like contesting, we have some QRP-only type contests. Most major contest like the **ARRL DX Contest,** the **CQ DX WW** and **WPX Contests** all have entries for QRP stations. The playing field is leveled in these contests since you'll only compete against other QRP stations.

Since many QRPers feel the need to take their gear with them almost everywhere they go, operation from the field has become extremely popular. The granddaddy of them all is, of course, the **ARRL Field Day**, the last full weekend each June. Many regional clubs sponsor to-the-field type operating events or sprints prior to and after **Field Day**. There is the **Freeze Your Butt Off (FYBO)** contest early in the spring where you get extra points for enduring extreme cold while operating from the bush. Then there is **QRP to the Field**, along with the various activities of the **HFPack** group whose goal is to spend lots of time in the bush or as pedestrian mobile/portable. Let's not forget the annual **Flight of the Bumble Bees**, or the **BUBBA** contests or the monthly **Spartan Sprint**, where the weight of your rig counts against your overall point score! All these operating events are field oriented providing many hours of fun in a mildly competitive, contest-like environment.

Each spring and fall, the QRP ARCI sponsors their **QRP QSO Parties**, which are weekend-long contests that attract thousands of QRPers. The interesting thing about these QRP-only contests is that the signal levels of your fellow QRP contesters are amazing. Sure, there are weak signals on during the contests, but for the most part, the QRP signals I have encountered are in the S-7 to S-9 + range, indicating that under most conditions, QPR is more than up to the task of providing reliable long-haul communications.

At the end of each year the G-QRP-Club sponsors their **Winter Sports** operating event. Starting on Boxing Day (the day after Christmas) and continuing through the New Year, it is a time for QRPers worldwide to get on the air for some wintertime fun.

HOMEBREWING

FAQ: I DON'T CONTEST AND I HATE THE IDEA OF GOING INTO THE BUSH WITH MY GEAR. WHAT OTHER INTERESTS ARE THERE TO PURSUE IN QRP?

Without a doubt, the most intriguing aspect of QRP is the ability to build small, low powered, easily transportable gear. Long thought to be near extinction, the art of homebrewing gear has made a tremendous comeback within ham radio thanks to the QRP movement.

Many of the regional clubs have small kit projects ranging from SWR/power meters and accessories to simple transistorized QRP rigs all the way up to and including digital transceiver kits. Whatever your level of electronics and fabrication expertise, there is something for you within QRP. The kits offered by the clubs are done on shoestring budgets, passing the savings along to the members. As an example, the New Jersey QRP Club's 80 meter Warbler, an 80 meter PSK31 transceiver kit costing around $50, has provided an entry point for many of us old timers into the world of digital ham radio. What a bargain! Fifty bucks for a functional transceiver kit that introduces a new mode and sharpens your kit building techniques by introducing the builder to surface mount technology. You can't go wrong there!

ANTENNA EXPERIMENTATION

When you give up 13 dB of RF power to play in the QRP arena, you must ensure the efficiency of your station is up to the task. The single most critical portion of your QRP station is the antenna. If you have an inefficient, poorly erected or designed antenna, you will become quickly frustrated trying to make contacts at the 5-W level. The inefficiencies of your antenna system won't be apparent at higher power levels simply because your effective radiated power (ERP) will not become as impacted at 100 W as it will at 5 W. It's when you drop down to QRP levels that the inefficient antenna system really becomes apparent.

Fortunately, the antenna system is the easiest thing to deal with in the low power game. Next to homebrewing gear, antenna experimentation is one of the most rewarding aspects of QRP. True, you don't have to engage in QRP to play the antenna game but QRPers are definitely on the forefront when it comes to designing, building and erecting efficient, high performance antennas.

Long time QRPer Roy Lewallen, W7EL, designed a program called EL-NEC which was a very easy to use antenna-modeling program that enabled QRPers to model and tweak their antenna systems prior to erecting them. Roy upgraded the program and re-released it as EZ-NEC, which is a full featured Windows ™ antenna-modeling program that has gained world-wide acceptance among antenna experimenters.

Through the use of antenna modeling, today's QRPer can erect some killer antennas that will definitely level the playing field on the bands. Simple wire antennas are, by far, the most prolific, the least costly and easiest to erect. There is nothing wrong with using wire antennas. Quads and Yagis are nice, but wires get the job done, with much less hassle and maintenance, not to mention neighborhood impact!

Nowhere but QRP has there been so much interest generated in portable and mobile antennas. The **HFPack** group sponsors an "antenna shoot-off" each year at the Pacificon ham radio conference. This event pits antenna makers, commercial and individual, to put their creations to the test on an antenna test range. The results are sometimes very surprising. Simple home-built antennas often outperform their commercial counterparts that cost well over $100!

VINTAGE QRP

Having cut my electronics "teeth" on vacuum tube equipment, the desire to assemble a vintage vacuum tube QRP station has been a long-time goal. I am not alone. There are many QRPers who yearn for the yesteryears of ham radio that feature the warm glow of tube filaments and the distinctive smell of warm electronic equipment.

One of the most popular topics I have encountered during the years of writing *QRP Power* column for *QST* has been vintage QRP gear. We're not just talking a simple one or two tube crystal controlled transmitter, either. I am talking about completely homebrew vacuum tube stations with regenerative or simple superhet receivers, multi-band transmitters, a Vibroplex Bug™ and Dow-Key™ antenna changeover relays! This is *REAL RADIO*!

Bill Albrant, K6CU, uses an ARC-5 transmitter and receiver as part of his portable Field Day station. **(Photo courtesy of K6CU.)**

The interest in "toob" gear has resulted in a newsgroup called the ***Glowbugs List*** on the Internet. The technical expertise available on this newsgroup is phenomenal. Old timers and no-so-Old Timers take the time to help newcomers into the fold of vacuum-tube electronics. Whether you have questions about a choke-input high-voltage power supply or help aligning a restored Hammarlund HQ-129 receiver, the folks on the ***Glowbugs List*** will be there to lend a hand.

In closing this chapter, let me say that QRP has finally emerged as main stream ham radio done at the 5-W (or less) level. The diversity of this facet of the hobby is rivaled only by the diversity of the people involved with the QRP Movement.

CHAPTER 2

The Advantages of QRP

It should be obvious by now that I am in love with the low power aspect of the ham radio hobby. Not only is QRP operating tremendously rewarding, challenging and fun, it is the "right thing to do" when it comes to life in the 21st Century.

For years ecologists have been telling us that we need to change our lifestyles and the way we do business here on Planet Earth. The planetary ecology is impacted by everything we do, or in some cases, don't do. The same can be said for ham radio. As more and more radio amateurs yield to the "dark side of the

***Mike Martell, N1HFX, is never happier than when he is operating a QRP station. If the station uses a rig he built, his smile is even bigger!* (Photo courtesy of N1HFX.)**

Force" and fire up high-powered stations, the noise and interference levels on the bands has risen to epidemic proportions. Arrogance reigns supreme among some of these QRO-types. Some of the "Watt Hogs" and "Ether Busters" won't even talk to a QRPer once they find out he or she is running only 5 W or less, although the QRPer's signals are very readable! Talk about rude!

One of the primary reasons for running QRP is to reduce the amount of interference on the bands. It makes good "RF Ecology" to limit your power output so RF pollution can be kept to a minimum and *everyone* can enjoy the hobby.

One of the newest threats to our radio hobby is Broadband over Power Lines (BPL) Internet service providers. While there are doomsayers predicting the ultimate demise of ham radio as we know it, I tend to think that BPL will be about as big of a fizzle as the predicted Y2K computer failures! If only $^1/_{10}$ of the dire predictions come true, however, HF communications will never be the same again. It won't just impact QRPers, either. No sir, the high power boys will also feel the strain of tons of RFI/EMI generated by the BPL equipment.

RF EXPOSURE CONCERNS ARE MINIMIZED

Right down the ecology alley is concern about RF exposure. In only the last few years, with the advent of so many portable cell phones, RF exposure has become a major concern to the radio and health-care industries. Any time you have a human body in close proximity to a large RF field, you run the risk of overexposure to RF energy. How much is too much? That depends upon the frequencies involved, your physical proximity to the radiating device, how long you are in the near field, etc. The ARRL has published *RF Exposure and You* with lots of information and tables to help you evaluate your station for safe operation. Both the Technician and General class question pools include questions and answers about RF exposure, so the ARRL license study materials include information about this important topic. Check out the information to be safe, and err on the conservative side.

With QRP, your RF exposure concerns are minimalized. It doesn't take a rocket scientist to figure out if you are running 1500 W on HF compared to 5 W, your exposure to hazardous RF energy is much less likely to occur using the 5-W station. In fact, stations running less than 50 W PEP on any frequency are exempt from performing the otherwise required RF Environmental Evaluation. The evaluation is intended to demonstrate that any people who are exposed to the RF radiation from a transmitter will receive less than the FCC maximum permissible exposure (MPE) on any given frequency. So QRPers won't have to perform that evaluation, according the the FCC Rules. Your station will still have to produce less RF exposure than the MPE limits, but as long as no one can *touch* your antennas while you are transmitting that should not be a problem.

OFF THE POWER GRID

QRP operation, and the associated small power consumption by the gear involved, dovetails nicely with using renewable energy sources. Photovoltaic cells (solar panels), wind and water power generation schemes all can be used to keep a modest QRP station on the air indefinitely with little or no impact on the ecology or the planet. Imagine, having fun using ham radio with non-polluting energy sources that are readily available at modest cost. What could be better? With the average cost of a solar panel at approximately \$5/watt, a 60-W panel would cost only \$300 (new). These panels are extremely rugged and the typical life of a

***Jay, KA1PQK (left) and Jim, K1PTF (right) Francis enjoyed camping and operating this simple QRP station as a team for the 1998 Field Day.* (Photo courtesy of K1PTF.)**

photovoltaic panel is around 20 years. Two 60-W panels will provide the average QRP station with enough power year round by charging a battery bank (normally two deep cycle RV/Marine batteries) at a cost of only $.082 per day over 20 years. Now, *that* is cheap! Add to this the fact that your station now is entirely independent of the commercial power grid, and you can remain on the air during emergencies when others are shut down. This brings into close focus the value of using a QRP station for emergency communications. We'll discuss emergency communications and QRP in a later chapter.

OTHER ADVANTAGES

Radio Frequency Interference and Electro Magnetic Interference (RFI/EMI) are on the rise across the RF spectrum. We have, in just the last few years, become a "wireless world." Seemingly *everyone* has a cell phone: the kids, mom and dad, the boss, the cops, everyone! There is a trend to "go wireless" with your Internet connectivity, using the IEEE 802.11 microwave infrastructure. Then, there is BPL! You get the picture. Lots of stuff to get into our QRP receivers.

What about us getting into them? It happens all the time. Not nearly as often with QRP power levels as with 100 to 1500-W stations, but it does happen. The most often affected personal convenience is the telephone. Even slight amounts of RF energy from a QRP rig can tear up any of the solid state phones. Lots and lots of ferrite cores is one answer.

Nothing endears you to your neighbors like tearing up their stereo, TV, VCR/DVD player, telephone and toaster! Yup, you can really make friends and influence people with a good dose of RFI/EMI when you're a ham radio operator. Needless to say, the occurrences of interference from QRP transmitters is much, much less than with the higher power variety. Therefore, once again, it makes good sense to drop the power back and play the QRP game. not only for the fun and the challenge, but to keep peace with your neighbors/landlord.

One other aspect of running QRP, especially in crowded urban areas, is that this mode is very "Condo Friendly." Many times condo dwellers find that they are not able to erect any form of antenna due to covenants and restrictions placed upon them by the condo committee. A lone no. 30 AWG wire snaked out an upstairs window to a nearby tree and suitable indoor counterpoise and a couple of watts will get you on the air from a condo or high-rise apartment in fine style and nobody will know you are an active ham radio operator. The lack of RFI/EMI, coupled with a nearly invisible antenna will make for many hours of fun and games on the bands.

QRP operation does not take a big bank account. Nor does it take a room full of communications gear. One of my Appalachian Trail radio sets occupies only a few cubic inches of backpack/butt pack space and allows single band operation on either 40, 30 or 20 Meters with 2-W output. Using wire antennas and a homebrew antenna tuner, I can put a good signal on any of the three bands. Cost of the radio sets: $80 each (they are three Wilderness Radio SST QRP rigs, one for each band). The antenna wire was stuff I had laying around and likewise the tuner parts came from a couple of transistors AM radios that I scavenged. The main coil in the tuner came from an Amidon source: $2.95. So, for a cost of well under $100, I have a very useable QRP rig that I can put on the air from the main station or the trail.

Used QRP gear fills the files of e-Bay (not that I would recommend this as a primary source of used gear), and hamfest flea

Mike Gaude, WK6O, is a picture of concentration as he operates his station during a backpack camping trip.
(Photo courtesy of WK6O.)

markets. Pre-owned Ten-Tec Argonauts (the analog versions) along with Heathkit HW-8s and 9s, all can be purchased for reasonable prices. Of course, there is the "collector mentality" out there that will go to extremes to obtain a relatively common piece of gear at an exorbitant price, thereby artificially inflating the costs of this gear for us mere mortals. A classic example of this is the Ten-Tec Argonaut 515. Less than 1000 of these last of the analog Argonauts made in the mid 1980s. They are really a very nice radio but they now go for almost as much used as they sold for new! In reality, this rig is worth around $300 to $350 but sellers can get well over $500 for a stock 515 with no accessories! Go figure.

While you can spend a bundle on a QRP rig, you can also use the rig you currently have by simply turning down the RF drive on CW to reach 5-W output. SSB can be a bit trickier, but by fooling the ALC circuitry, you can work QRP SSB at no further expenditure of cash. That way, you can get your feet wet without investing a bundle of money. Should, for some unfathomable reason, you not like QRP, then you can easily bail out and find some other ham radio interest without fear of having to off load a QRP rig to recoup some of your cash outlay.

One thing that giving up 13 dB of power will definitely do is force you to become a much better operator. Without the cushion of additional RF output, you are reliant upon your skills as a radio operator to be successful with QRP. So, another reason to pursue QRP is to sharpen your operating skills. An experienced QRP operator is an amazing individual. He or she can seemingly pull call signs out of the noise, snag rare DX with only a couple of calls, muscle their way into a pile-up and exit with the DX operator's call in the log, work contests like a machine, and so on. All this is done over time, by dedicating oneself to the task at hand: namely honing the operating skills to perfection.

Finally, QRPers are at the center of developing cutting-edge technology within the ham radio hobby. PSK31 comes immediately to mind. This mode is absolutely made for low power. Originally developed as a possible replacement for radioteletype (RTTY), PSK31 offers extremely small bandwidth utilization along with almost error-free digital keyboard-to-keyboard communications. Small Wonder Labs offers several complete PSK31 transceiver kits for different bands. This mode has picked up a

Dave, N6JKR (left) and Mike, WK6O (right) Gaude enjoy the fun of QRP operating during their 1998 Field Day backpack camping trip. **(Photo courtesy of WK6O.)**

huge following both in the QRP and QRO worlds. The proponents are a diverse group who enjoy rag chewing, DXing and even contesting using PSK31.

As you can plainly see, pursuing the art and science of QRP can be tremendously rewarding as well as loads of fun. The challenge of using 5 W or less on today's bands really separates the men from the boys, so to speak. Without the cushion of raw RF power, successful QRPers are forced to develop and hone their operating skills well beyond those of the average ham radio operator. By operating at QRP power levels you not only help reduce RFI/EMI, you expand you own horizons as a top notch operator, further adding to your enjoyment of the hobby.

CHAPTER 3

Getting Started

There are two very different and distinctive ways to start into the QRP hobby. The first is to go out and buy a new or used QRP rig or kit, get it on the air and play radio. Alternatively, the easiest thing to do is simply turn down your RF drive for 5-W output on CW and get on the air using the rig you normally use for HF work. The latter is the most attractive to most hams, since not only is it the cheapest, you get to use a radio whose operating characteristics you are intimately familiar with.

Cranking back the power on a 100-W radio is as simple as reducing the drive in the CW mode, until the full carrier output of the radio shows 5-W output on an in-line wattmeter. Elegantly simple. Most of today's rigs will accommodate 5-W output at the lower end of their drive range without any internal modifications or readjustments. If you encounter a problem, however, consult the manufacturer or look on the Internet and there will definitely be some advice available to help you get the RF backed down to QRP power levels.

SSB is a slightly more complex matter. Many times, you will need to provide some kind of adjustable battery bias to the ALC circuitry (normally a jack on the back of the radio). Alinco and Icom HF rigs are famous for this. It is just a matter of putting a 9-V transistor-radio battery and a variable resistor into a small out-board box and adding a cable to plug it into the ALC jack on the back of the rig. See **Figure 3-1**. Varying the battery bias will reduce the RF output to below 5 W on these types of radios without resorting to internal modifications or readjustments of the internal circuitry.

If you *must* spend money to "enjoy" the game, there are a

Figure 3-1 — This schematic diagram shows an ALC bias circuit that may be used with some commercial rigs to reduce the power output.

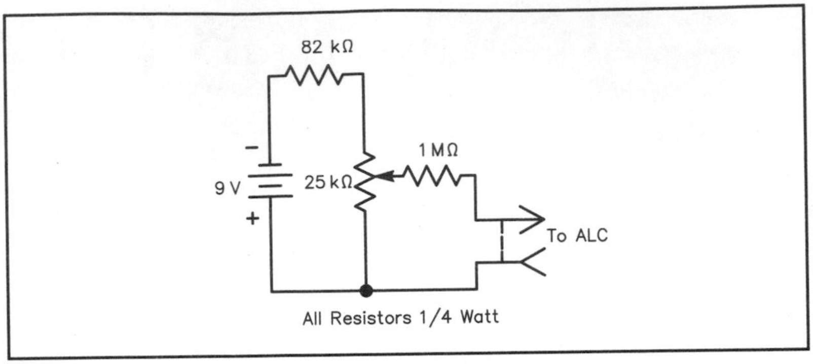

number of options open to help you lighten your wallet. First, of course, is to buy a new commercially manufactured QRP rig. In just the last several years the commercial market has opened up tremendously for the ham wanting a quality commercial QRP rig. Yaesu started it with their FT-817 and ICOM followed suit with their IC-703 and 703 Plus radios.

One thing to keep in mind is that any time you are interested in purchasing a new piece of gear, consult back issues of *QST* or go on *ARRLWeb* (**www.arrl.org**) and go to the product reviews for an in-depth look at the gear of interest. Nobody does product reviews better than the folks at the ARRL Lab. Ed Hare, W1RFI, and the Lab Crew take great pains in presenting factual technical information regarding the new equipment that enters the ham radio market. Their lab reports, coupled with a narrative report from a *QST* staffer, makes for interesting and informative reading that will assist you in making an informed decision on your next piece of gear.

One other comment: shop around using the 800 numbers in the ad section of *QST*. Many times the commercial manufacturers will offer incentives on selected pieces of gear during certain times of the year. This is your chance to save big bucks when buying a new rig.

PSSSSSST....HEY, BUDDY. WANNA BUY A QRP RIG?

Vertex/Standard (Yaesu) started the shift in commercial ham radio marketing towards the QRP market with their FT-817, probably the most successful new product to hit the streets in the last

25 years! Tens of thousands of these little portable multi-mode QRP radios are on the air from all over the world. There are at least two FT-817 groups on the Internet that have a tremendous source of information available to the new owner (check out **groups.yahoo.com/group/FT817/** and **groups.yahoo.com/group/FT-817/**). Prices vary, but you can pick up a new FT-817 for under $550 and a used one will cost somewhere around $400 and up, depending upon options and accessories. This is a great little rig to start with, since it covers all the HF bands (160 to 10 meters) plus 6 and 2 meters *and* 70 cm. Modes: CW/AM/SSB/DATA and it is 9600 baud packet ready! Check out their website at: **www.vxstdusa.com**.

Not to be left behind, ICOM recently marketed their IC-703 (this is the Japanese in-country 10-W version), which can easily be backed down to 5 W. This was followed by the IC-703 Plus (HF plus 6 meters). Both radios have built in automatic antenna tuners, which allow easy set up and operation during camping trips and outings. These are fairly new radios, so the street prices will vary and the used prices will be fairly high until the used market gets up to speed. Patterned after their very successful IC-706 MkIIG, the IC703 (Plus) is a solid performer at a great price. The addition of 6 meters on the "Plus" version is an excellent decision, since a few watts on six will allow you to work more than your fair share of DX when the band is open. The IC-703 (Plus) has a lot of bells and whistles and is at home in the shack as well as in the bush for Field Day or in support of emergency communications.

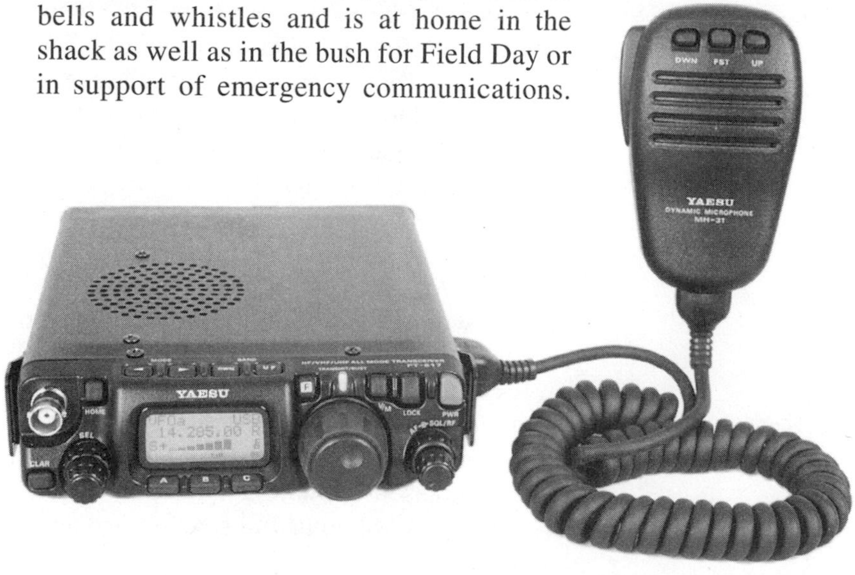

Point your browser to: **www.icomamerica.com** and take a look at their product line.

While Kenwood does not offer a "QRP" radio at present, their TS-50 has been used by QRPers quite successfully in that role. The RF can be reduced to 5 W and this rig is a good performer. On the Internet go to: **www.kenwood.net** for their latest products.

Of course, no discussion of commercial QRP rigs would be complete without a mention of the folks at Sevierville, Tennessee. Ten-Tec has their new Argonaut V available. This is a departure from their previous attempts at a new generation radio. This newest Argo is software upgradeable, so theoretically, you'll never have an obsolete rig. Prices are from the factory, since there are no longer any Ten-Tec authorized dealers. Don't forget to check out their used gear for a bargain on a factory refurbished Argonaut V. Try **www.tentec.com** for a look at their products.

If you are adventurous and want to have a lot of fun and learn something in the process why not try your hand a building your QRP rig? There are a host of kits for various budgets available for the enterprising QRPer. The obvious advantage to building your own rig is that you save money. The other spin-offs include broadening your electronics skills and the intense pride that goes with the knowledge that you are using a rig made by your own hands to communicate on a global scale.

The frugal QRPer can purchase a single band CW rig for as low as $25 (the SWL RockMite)! Conversely, if you want to go all out and get that dream rig, Elecraft offers their K2 with enough options to keep you kit building for several months, if not years! There is a huge difference in prices, but the gamut of what is available is open to almost any type of budget.

The premier QRP kits are from Elecraft (**www.elecraft.com**): the K1 and K2 radios are top of the line, first-class construction projects that result in extremely useful radios for the QRPer. Prices for the K1 start at around $350 for the four-band model without any options, and prices for the K2 start at around $679 for a CW only radio with no options. While the K1 will always be a CW only rig, the K2 can be expanded, via options, to include SSB and data. Elecraft just released three new transverters for 6 and 2 meters and 70 centimeters that will work on any transceiver with a 10-meter output. When used with the K2, all bandswitching and frequency displays will be done on the

K2, resulting in a very nice HF-UHF terrestrial weak-signal station that even the most ardent VHFer would be proud to own.

Dave Benson, K1SWL, from Small Wonder Labs (**www.smallwonderlabs.com**), offers a great selection of mid-range QRP kits. His SW-XX+ rigs are great little single banders for around $50. A lot has been written about optimizing these rigs, so once you get some experience using the stock radio; why not delve into the black art of modification? The SW-XX rigs offer a great little RF platform to play around on. Dave also produces a series of PSK31 transceiver kits designed to get hams interested in HF digital communications. Not only do you get to build the rig but you broaden your experiences using a new communications medium!

"QRP Bob" Dyer, K6KK, from Wilderness Radio (**www.fix.net/~jparker/wild.html**) offers several kits that cover the basic single band CW radio up to a multi-band QRP rig. His NC-40A, the commercial version of the NorCal-40 transceiver kit, has sold like hotcakes and continues to be a solid performer on 40 meters. Some folks have modified the NC-40A to work on 30 and 20 meters with good results. Cal-Tech uses the NC-40 as an undergraduate project for aspiring electronics engineers as part of their curriculum. Talk about success!

Wilderness Radio also markets the SST, a scaled down, minimalist version of the NC-40A. These tiny single-bander kits are ready to go in about 3 or 4 hours on 40, 30 or 20 meters. They are a really fun little rig and are great for backpacking and business trips, where you need to kill some time without becoming a slave to the boob-toob.

Finally, the flagship of the Wilderness Line is the Sierra, a Wayne Burdick, N6KR, design that has appeared in several issues of the ARRL Handbook, starting in 1995. This multi-band CW rig is really sweet, featuring band modules that you swap out to change bands. This reduces the T/R changeover circuitry, does away with noisy PIN diodes, and allows the user to buy and build only the modules needed for the type of operations desired. This rig comes with an analog front panel that can be upgraded (by the optional KC-2 keyer/digi readout) to a digital-read-out rig with internal memory keyer. The Sierra holds the world's record for QRP distances worked during a two way contact on 40 meters. That record was set on December 26, 1994 (1.9 ***MILLION*** miles per watt) and again in December 1995 (4 ***MILLION*** miles per watt) by Fran Slavinski, KA3WTF and Paul Stroud, AA4XX. (There is more about this record in Chapter 9.)

Oak Hills Research offers a couple of outstanding kits for anyone interested in"rolling their own". The OHR-100 is a single bander for 40, 30, 20 or 17 meters, and the OHR-500 is the five band version. These are CW only rigs, with good receivers featuring crystal lattice filters. Options are available for an internal CW keyer and an external digital dial. There is a lot of extra space inside these rigs to allow for ease of building and later modifications/options. They are straight forward QRP rigs that offer great performance at a reasonable price.

There are other kit manufacturers out there, such as Red Hot Radio, Ramsey Electronics, MFJ Enterprises (**www.mfjenterprises.com**) and Vectronics (**www.vectronics.com**) that offer various QRP rigs and accessories. In short, should you desire to build all or part of your QRP station, you can find a rig for the right price that fits your level of electronics fabrication skill and budget.

ANTENNAS, THE GREAT EQUALIZER

Okay, you've decided on a rig. The next *big* topic that needs addressing is what antenna(s) to use. As with the radio gear, use what you have. In an earlier chapter I eluded to the fact that wire antennas were still the mainstay of ham radio. Nothing beats a simple half-wave wire dipole up in the air at a convenient height. Actually, closer to the ground might be better initially, since it opens the door to localized communications that you cannot obtain with the smaller take-off angles associated with dipoles raised ¼ wavelength or higher above ground. This allows you to get your feet wet with local contacts (within a couple of hundred miles of your QTH) and experience the thrill of QRP before trying your hand at DXing.

Obviously, if you have a triband beam or quad on a tower then you have the basics of a good QRP DX station. The directional properties of the rotatable gain antennas really tend to level the playing field. This is a striking demonstration of the ability of antenna efficiency offsetting the 13 dB power disparity associated with dropping your RF output down from 100 W to only 5 W. Believe it or not, I have been in QSO with operators who did not believe that I was running only 5 W into my TH7DX beam. It is not unusual to get S-9 +10dB reports using a good beam or quad antenna while only running QRP power levels!

Some QRP purists would say that this is not fair and definitely not in keeping with the spirit of QRP. Oh, pooh! Efficiency is the name of the game. Our definition of QRP deals with the output power levels, *not* effective radiated power (ERP). If you want to live in a cave, make your rigs from bailing wire and snoose can lids, and power it all by a hamster-driven exercise wheel, fine, go ahead! This hobby is big enough for all of us. I, for one, will applaud your efforts and laude your results! Just don't expect me to give up an efficient gain antenna system because you don't think it is in keeping with some obscure idea that in order to participate in QRP operation you must endure hardships that would try the patience of Job. QRP is *F-U-N!* When it ceases to be so, it's time to find another hobby!

Probably the one QRP station accessory that you should definitely have is a QRP wattmeter or SWR bridge. Oak Hills Research makes a wonderful one evening kit: the WM-2 QRP wattmeter. Costing around $90, this simple accessory can take the mystery out of setting your power output and checking out the reflected power on the antenna feed line. The WM-2 has three full scale readings: 0 to 100 mW, 0 to 1 W, and 0 to 10W. It offers a reverse detection mode for measuring reflected power. Housed in an attractive gray cabinet with a large custom meter, the WM-2 is *the* QRP accessory I would recommend for anyone in the hobby. The North Georgia QRP Club (**www.nogagrp.org**) offers kits for a QRP SWR/wattmeter. The NoGaWatt project is a variation of the "Stockton Wattmeter" that can measure both forward and reflected power simultaneously with dual meters. This is a handy feature.

BASIC QRP OPERATING TECHNIQUES

Alright, enough of this stalling. Let's do this the easy way and tune up your high power rig on 40 meters and load the antenna for lowest SWR/reflected power. Now, terminate the radio into a dummy load and back down the RF drive in the CW mode to where the meter registers 5 W key down. Okay, we're set. Time to become a QRPer.

The first question that normally comes up at this time is whether or not to call "CQ". Realistically, your 5 Watt signal should radiate just fine and calling "CQ" should pose no major problem. In the real world, however, your 13 dB sacrifice of RF power will lower your signal slightly over 2 S-units and that may be enough to have stations tune right over your signal when you're calling "CQ". So just to be on the safe side, let's hold off on the "CQ" thing for a couple of contacts and slide down around the "40 meter QRP Watering Hole" at 7.040 MHz. Here is where you'll find other QRPers waiting in the weeds to have a chat.

Listening is about listening. Duh! Listening is the single most important skill (next to patience) that a QRPer must learn to develop. I cannot stress this enough. Good listening habits are the key to success with any weak-signal work in general and QRP in particular. The only way you develop good listening skills is by spending time in front of the radio. There is no substitute for T-O-R (Time-On-Radio). Not only do you learn how to listen you'll learn how to interpret band conditions, a vital skill that can spell success or failure in QRP. We'll discuss "band scanning" in another chapter, but don't let fear hold you back. Start carefully listening to the bands now and you'll be one step ahead when we get to band scanning later.

Let's just hang out here on "040" and see who's home. Without a doubt, 40 meters is the most popular QRP band and you'll be surprised at who you will run into there.

The most important part of the listening game is to learn to tune through the frequencies slowly...spelled: s-l-o-w-l-y. There is a tendency in QRP neophytes to tune the band rapidly. This is a sure fire way to miss contacts. Slow down! Now! You'll often find one or more QRP stations clustered about one frequency and you'll need to employ some fine tuning techniques, coupled with judicious use of the on board filtering and/or DSP to pull a QSO out of the mud. The other thing about developing your listening skills is that it forces you to learn how to *really* use your transceiver's functions and associated accessories. You would be surprised at how many hams don't know how to effectively use their multi-thousand dollar radios!

Okay, start tuning. Slowly...slower...that's better. You're coming up on a signal, so carefully line it up on the receiver passband. Good. Let's sit and listen. It's a VE3 in contact with a 4-land station, both of them running QRP. Notice the slow fade on the VE3's signal. Also notice that both of these QRP signals are averaging about S-5, very readable signals. Notice, too, that the CW speed is not all that fast. Normally, QRPers are not speed demons and you'll find the average CW speed in a QSO is around 15 to 18 wpm.

Okay, let's move the VFO down a bit and continue slowly tuning. Oops! Hold it! There is someone cranking along about 40 wpm. Don't worry, I can't copy him either. Nice signal, though. A solid S-9 +. He's signing it back to the other station. I copied "XX" as the last part of his call, just before he stopped. There's his contact, a W2, nice and loud. Let's see if we can grab the entire callsign. "AA4XX," Paul Stroud near Raleigh, North Carolina. I should have known by the flawless CW and the outstanding signal. It should come as no surprise that Paul's three element 40 meter wire beam at 60 feet puts out a blasting signal. I just wish I could copy CW as well as Paul. He's an outstanding operator.

Okay, let's camp out here and try to improve our CW skills by listening in on Paul's QSO. Sure, neither of us is going to copy much of the conversation, but it never hurts to push the envelope, so to speak. By trying to copy well above our comfort level, our CW skills will improve. Besides, once AA4XX finishes he'll most likely hang around this frequency and we'll give him a call.

Now, let's discuss the best way to get the attention of the other operators as they finish up a QSO. You could do the old standard of a three-by-three call: his call repeated three times followed by "de" and then your call three times. This is a bit slow, and really unnecessary.

A better way would be to "tail-end" the W2 station, the one who's in QSO with Paul, just as he signs it over to Paul on the last go-round. Tail-ending amounts to sliding your call in just as the other station signs off. Since your call is going to be lower in signal level, you're not going to interfere with the other station to any extent, and, with a first class operator like Paul, he will definitely hear your call sign in the background. Simple, slick and quick!

All right, the W2 is starting to sign, get ready! Hit it! Once and only once: just your callsign. Now we wait. Whadda know? Paul is now calling little old you! Congratulations, you just initiated your very first QRP contact and it's a two way QRP QSO to boot! Now, talk to Paul, he'll slow down, I promise.

Way cool, dude! Well, you should feel pretty good about now, with your first QRP QSO under your belt. Paul gave you a

599, which is a fine signal report. Of course, you're looking right down the throat of his huge 40-meter beam, but hey you just logged number one!

Okay, let's continue to tune around 7040. Notice that there are some SSB signals in and around this frequency. That is from our Canadian brothers whose phone band is slightly lower in the band than ours. Hey, no problem, we just get creative with the IF SHIFT and DSP controls on the receiver. They won't bother us and I doubt we'll bother them.

Hmmmm, everyone is in QSO around the QRP "watering hole" so let's slide down the band a bit. Slowly…I said *slowly*. That's better. Make that conscious effort to slow your tuning down so you won't miss other stations on the band. This is especially critical while DXing, when the DX station is weak and right in the noise. What's that? A nice strong CW signal doing about 20 wpm and he's calling "CQ". Go ahead, as soon as he signs, fire off your callsign, *only once*. Bingo! He's coming back to you. Go ahead, get busy and work him.

He's a W3 in Maryland and he just gave you a 559 signal report. Not bad, so swap the pleasantries and give him an honest report: I'd say 589 based upon your S-meter deflection.

About this time you may wonder when it is the proper time to tell the other station that you are QRP. Based upon countless contacts, I normally wait until the third or fourth go-round and then drop the bomb. Why? Simply because if the other station can copy your QRP signals over a period of time, then you know that his operating skills are good, he hasn't fudged the signal report, and he can handle the news that you're only running 5 watts. Sometimes telling the other station that you are QRP can backfire. All QRPers have had incidents where the other station, upon finding out you are QRP, will suddenly have problems hearing your signals and bluntly drop the QSO, stating that unless you have a linear to "turn up the wick" he doesn't want to talk to you! What a pity. It's the way some high-power ops deal with us QRPers.

Thankfully most of the non-QRP contacts we make don't end that way and the other station often applauds our efforts and starts asking questions about running low power. I have gained "QRP converts" this way. Once the other station sees that your 5-W signal is not difficult to copy, and the QSO is not a fluke, the probability of him becoming interested in QRP greatly increases.

So far you have bagged two QSOs running only 5-W output. Continue tuning around 40 meters and see what else you can dig up. Once you initially break the ice on using QRP, you'll find that 5 W is more than enough power to work the majority of stations you can easily hear on the bands. Once you transition to this phase of the hobby, you will rapidly gain confidence as you work more and more stations. Confidence is what it takes to ultimately tackle DXing and contesting while using low power.

To recap briefly, what we tried to do in this chapter is to show you, using statistics and a couple of imaginary QSOs (actually taken from real life Qs I've had), that getting on the air with QRP is not only relatively easy to accomplish, but you can actually make contacts at these power levels. Once you break through the initial reluctance to try QRP, you'll find that low power QSOs come quickly and you'll gain valuable experience as a neophyte QRPer by just getting on the air. The other thing I want to stress at this juncture is that the signal strengths encountered when working another QRP station will vary, but for the most part, they will be quite readable. This is contradictory to established thinking, which says that "the other guy" does all the work when working a QRP station. My experiences, along with those of the main-stream QRPers around the world, indicate that a 5-W signal is relatively easy copy under all but adverse band conditions.

CHAPTER 4

QRP Equipment

The one area of QRP that has changed dramatically is the commercial and kit-based transceiver market. The commercial QRP equipment market is booming and QRPers all over the world are reaping the benefits of commercially manufactured QRP equipment with outstanding features, ease of operation and portability.

On the commercial scene, SGC revamped their stalwart 2020 transceiver by improving the receiver and adding audio-derived DSP. Yaesu threw their hat into the ring by offering the FT-817. If sales records are any indication, the FT-817 is *the* QRP radio of the New Millennium. ICOM also recently released their IC-703 +, a nifty little QRP version of their hugely successful IC-706. Ten-Tec has also gotten back into the QRP arena by offering their new Argonaut V to complement a long lineage of solid-state QRP transceivers going back to the early 1970s.

Lots of kits have hit the QRP scene in the last few years. Some of these kits are still around, proving their worth by performance under fire, while others never left the prototype stages or have fallen due to substandard performance. Various QRP Club kits have provided us with everything from the ultra simple to state-of-the-art rigs that use surface-mount components. They are fun to build, can be a challenge to use, and some are even full-blown digital QRP transceivers. There is no shortage of QRP rigs, that's for sure. There is something to fit everyone's budget. QRP does not have to be expensive!

This chapter is the longest in the book, simply because I want to showcase as many rigs as possible, giving my thoughts on each one that I have personally used. If your favorite rig is not listed, I'll apologize in advance, since there is only so much room in the book. I have tried to present equipment that is considered "main stream" in the QRP market, with universal appeal. At this point I must state that I do not own stock in any company that manufactures ham gear, nor do I get my gear at a discount price just to say nice things about the equipment in print. Finally, I have no monetary involvement with any commercial or kit equipment manufacturer. What I present here are my own opinions based upon actual use of the gear and upon published product reviews and lab specification reports. If you don't agree, that's fine. You're entitled.

I don't intend to dwell upon the QRP Club offerings except for the Tuna-Tin II transceiver that was updated in 1999 and offered first by NorCal, then the NJ QRP Club and currently by the Ft. Smith QRP Group. This kit is truly a piece of QRP history that you can own for the paltry sum of around $10.

My personal philosophy regarding test data is that while it is interesting, empirical test data does not tell the entire story. I know that there will be those who read this statement and vehemently disagree with my comments. From my perspective, however, test data is only part of the equation. On-the-air testing and operator skill ultimately win out over sterile test data every time. By combining lab test data and on-the-air testing (using a skilled operator) a very accurate picture of a rig's performance can be obtained.

This chapter is organized by featuring the currently available commercial transceivers and proven kit offerings together, listed in alphabetical order. At the end of the chapter there is a brief synopsis of the older used equipment that is still readily available on the pre-owned market. Without further ado, let's jump into the fascinating world of QRP transceivers.

ELECRAFT K2

In the first edition of ***Low Power Communications,*** I briefly described the K2, having seen only pictures and hearing only factory hype. The rig received high marks from postings on the Elecraft website reflector (**www.elecraft.com**). Initial indications were that the K2 was a red-hot performer. I remained somewhat skeptical, however, since my overall experience with kit radio performance led me to believe that kits were a series of compromises pitting cost against performance, with cost wining and performance suffering in almost every case. Any reservations I had regarding the K2 vanished after reading the March 2000 *QST*, where the first K2 review appeared. Subsequently, after building my own K2 (s/n: 971), I experienced first hand, the joys of owning one of the world's premier amateur transceivers. See **Figure 4-1**.

The ARRL Lab test reports presented in that *QST* review were nothing short of unbelievable. The K2 scored exceptional marks on receiver performance, leading ARRL Lab Supervisor Ed Hare, W1RFI, to publicly state that "Where the K2 really shines is in its receiver performance. On average, transceivers positioned in the upper tiers of the popular HF product lines (in the $2000 to $3500 price class) exhibit blocking dynamic range measurements somewhere in the vicinity of 130 dB and a two-tone, third-order dynamic range near 95 dB. The K2's receiver performance compares very favorably to that of the samples of the high-end radios we've recently examined, turning in impressive 136/97 dB figures for these parameters." That statement is worth re-reading. Imagine a kit radio costing under $600, whose receiver can run head-to-head with the Yasue FT-1000D, the Kenwood TS-870S and the ICOM IC-756! Now, that *is* something to write home about!

Timing is everything, as demonstrated by the phenomenal success of the Elecraft K2 transceiver kit. I seriously doubt that the

Figure 4-1 — The Elecraft K2 CW/SSB transceiver is a complete kit designed by Wayne Burdick, N6KR. This microprocessor-controlled radio has 10 memories, allows direct frequency entry from the keypad and has a digital display. Split frequency operation with dual VFOs as well as many other bells and whistles make this kit a very attractive piece of equipment. (Elecraft Photo)

K2 would have been such a success had it been introduced five years previously. The timing was not there. Starting in 1994, the QRP community has been treated to an abundance of outstanding small, high-performance rigs designed by some really talented engineers, all of which set the stage for the K2.

Wayne Burdick, N6KR (of Wilderness Radio fame) teamed up with Eric Swartz, WB6HHQ (We don't know where he came from but we're sure glad he's here!) and decided to collaborate on a CW/SSB transceiver kit. They called their new venture Elecraft. Their mission: to design, prototype, manufacture and market the world's most advanced QRP transceiver kit. The result: the K2. I can say unequivocally that they have achieved their mission goals.

We knew that Elecraft was a "different" kind of company. From the start, Elecraft solicited comments from the QRP community on what was really needed in a top-of-the-line QRP radio. They created a team of "Beta Testers" that took the initial one hundred K2 kits, built and debugged them, and provided valuable feedback prior to the company ever selling a single K2 kit to the public. This resulted in ongoing production and marketing delays but the wait, while exceedingly frustrating, was well worth it. Elecraft's direct interaction with the Beta Test Team produced a much improved kit that was virtually free of bugs and guaranteed to work the first time, provided the builder followed the instruction manual precisely.

My own K2 arrived in March of 2000. I completed the construction in about 30 hours. Outside of a couple of errors directly attributed to "operator headspace" the rig went together with no hassles. In order to purchase my K2, I sold my trusty Ten-Tec Omni-C, which was the best radio I had ever owned up to that point. I was quite apprehensive about selling off my favorite rig in order to buy the K2, but after building and aligning the kit, I've never looked back. The K2 outperforms any radio I've ever owned. It has accompanied me on a couple of trips and performance has been flawless for both casual operation and contesting at home and on the road. Only a couple of problems have arisen. I traced both to cold solder joints. In all instances, technical help is just an e-mail away. Elecraft maintains an active presence on their reflector. Any questions or problems you might encounter are quickly answered by the factory technical support staff.

Operationally, the K2's receiver performance is amazing. I have been able to get within 300 to 400 Hz of "Big Gun" contesters and, with the IF filters cinched down, still copy and work much weaker stations. That is outstanding dynamic range! There are four filter selections available for SSB, CW and now even RTTY. Dual VFOs plus RIT and XIT give the K2 extreme frequency agility. The keyer speed and RF power output are displayed as their respective controls are rotated. Operational settings can be modified on the fly via the **MENU** button.

Now a word about the K2's crystal filters. Filter bandwidth and center frequency are fully programmable via onboard software. Although the manual goes into great detail about how to do this, the easiest way is to use a shareware program called Spectrogram (downloadable on the web: **www.visualizationsoftware.com/gram.html**) and the sound card in your computer. After struggling for 3 weeks to properly align my K2's filters, Dave Carey, N3PBV, and I did the job using Spectrogram in about 10 minutes. The archives on the Elecraft List have all the details on this procedure.

Are there things I don't like about the K2? Sure. As with any radio, the K2 has some quirks. First is the QSK. Ten-Tec break-in is a lot better thanks to the small tuning range of their pre-mix VFO, which provides extremely fast TX/RX shifts. Elecraft opted for a non-premix VFO with wider tuning range, offering nearly general coverage and eliminating birdies. Ten-Tec rigs can switch between TX and RX in less than 10 ms. The K2 takes up to 20 ms. The K2 VFO also uses far fewer parts, and is easy to align...important features for a kit. Then there are the erratic power output readings. In comparing output readings obtained using a precision wattmeter terminated into a 50 ohm nonreactive load and the onboard RF metering in the K2, there were significant differences (on the order of 200 to 500 mW) each time the K2 was put in the "tune" position. Elecraft says that this is a result of antenna loading and the installation of the automatic antenna tuner should reduce the errors. In order to insure that you operate within QRP power levels, I highly recommend the use of an external wattmeter.

Do I like the K2? **ABSOLUTELY!** This is one terrific radio! The price of a basic kit ($599) is slightly more than one would expect to pay for a used QRP rig. By adding some of the more popular options to provide SSB operation, noise blanker, internal antenna tuner, internal audio filtering/real time clock, DSP, etc, you will have close to $1200 invested in the K2. This cost is offset by the superior receiver performance of the radio, which rivals the multi-thousand dollar commercial rigs! In addition, with the K2 you feel the pride of building your own gear.

THE ELECRAFT K1

With the introduction of the K2, Elecraft captured a large chunk of the QRP equipment market. Wayne Burdick, N6KR, and Eric Swartz, WA6HHQ, wanted to provide a high quality, feature-filled kit for the QRPer, which would be at home in the bush as well as the home station. The result: the K1, a compact, dual or quad-band CW transceiver featuring many of the bells and whistles found on its big brother, the K2.

This kit can be undertaken by anyone with some kit building experience under their belt. I would not recommend the K1 kit for a novice home constructor. If you have completed a couple of simple construction projects, however, you are ready for the K1.

The K1 is ordered in one of two configurations. First, there is the dual-band version with a choice of band modules for any two bands, in any combination from 80 meters through 10 meters. These dual-band modules can be swapped to provide true multi-band coverage. It takes a few minutes, but it can be done in the field. If you want the four-band version, however, it is available with provisions for 80, 40, 30, 20, 15 or 17 meters. When I ordered my original dual-band K1, I specified 40 and 20 meters because they give the best daytime/nighttime performance on a regular basis.

My K1 kit (s/n 017) arrived in the familiar white Elecraft box. The parts were fully prepackaged, and inventory was a snap. The manual is extremely high quality and guides the builder through every step of assembly and checkout. If you do get stumped, help is just a few keystrokes away on the Elecraft website. Construction took about 14 hours. During the checkout and alignment, I found several cold solder connections that were quickly remedied. Alignment proceeded without incident.

The beauty of the K1 transceiver is its portability coupled with multi-band operation in a very compact package (2.2" H × 5.2" W × 5.6" D). Total weight on my K1 with no options is about 2 pounds. Power output on the K1 can be adjusted from 100 mW to 5 W, and higher. Current consumption is around 60 mA on receive and 1 A on transmit, at the 5-W level. See **Figure 4-2**.

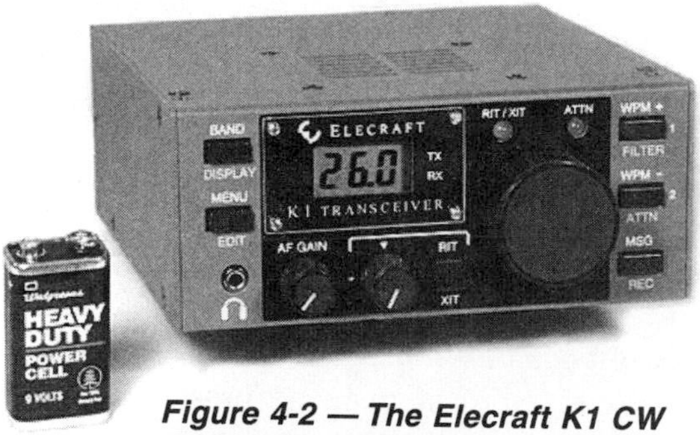

Figure 4-2 — The Elecraft K1 CW transceiver is built for either 2 or 4 bands. With an internal battery, internal keyer and an optional internal automatic antenna tuner it is a perfect station for backpacking or other portable operations where size and weight are important considerations.

This tiny rig has a lot of microprocessor circuitry inside. The multi-functional LCD readout displays the frequency (in megahertz, with resolution down to 100 hertz), S-meter readings, power output, keyer speed and dc supply voltage. The receiver features a single conversion superhet design with three software-selectable IF crystal filters. Like the K2, the K1's IF filters are adjustable. The transmitter circuitry is straightforward with a 2SC216 driving a 2SC1969 PA. Keying is clean and almost full break-in (QSK). The internal memory keyer will handle speeds up to 50 wpm. The keyer provides two programmable messages.

The tuning range of the K1 is selected during construction. There are two capacitors furnished that will yield a tuning range of 80 or 120 kHz. I tried substituting my own 75-pF silver mica cap in the VFO for a 100-kHz tuning range. After using this for several months, I replaced it with a 50-pF silver mica cap and achieved a 62-kHz tuning range. This yields slower tuning and better control of the ten-turn pot used in the VFO. My other mod consisted of changing C31 (AGC time constant) from a 2.2-µF to a 0.47-µF electrolytic cap. This drastically improves the AGC action but doesn't produce any irritating "thump" when keying.

My K1 has been on several trips to Florida. Operation to date has been flawless. The receiver is quite good in a high-RF-stress environment, allowing me to pick weaker stations out of the cacophony of noise on a crowded band. RIT and XIT give the K1 some frequency agility. Unfortunately, there is no way to split the transmit and receive frequencies more than a couple of kilohertz to allow for the wider split frequency operation favored by DXpeditions. The ability to accurately control the RF output down to 100 mW allows exploration of the challenging world of milliwatt ham radio. Since I already owned a K2, learning to use the "tap and hold" techniques to access the various features provided by the K1's microprocessor was a snap.

There are some optional accessories available for the K1 that make it a stand-alone QRP station. First is the auto antenna tuner that fits inside the rig's case. This is a scaled-down version of the very successful K2 auto-tuner. Also, there is a noise blanker and a nifty little stand that allows the operator to position the K1 in various ways for ease of operation. The stand folds flat for storage and transport. All you need to get on the air is a 12 V dc power source, antenna and keyer paddles. Elecraft also markets an internal battery option for the K1, which makes this a truly portable QRP station.

For those times when you need a small, highly portable rig to take on business/camping trips or to augment your shack, you couldn't do much better than the Elecraft K1.

THE ELECRAFT KX1

This is the transceiver that Wayne Burdick, N6KR, has always wanted to design and mass market. Billed as "the featherweight champ" the KX1 is a trail-friendly radio design with all the controls on the top of the case for ease of operation in the bush, at trailside, on a beach chair, in your sleeping bag or on a picnic table. Overall dimensions are 1.3" H × 5.3" W × 3" D. The basic KX1 covers 40 and 20 meters but there is an option to add 30 m to the mix, making the KX1 a three-band ultra-portable CW rig. The rig features a superhet receiver with adjustable crystal filtering, audio CW frequency readout of your operating frequency, and an LED logbook lamp. An internal battery provides 20 to 30 hours of causal operation. Options include the KXPD1 paddle set and a KXAT1 auto antenna tuner.

Not having the resources to build and review the kit for this book, I have asked Cam Hartford, N6GA to add his comments regarding the construction, alignment and use of the KX1 to this portion of the book. Here is Cam's report.

More than twenty years have passed since the publication of Roy (W7EL) Lewallen's article describing his Optimized QRP Transceiver (*QST*, August 1980). Roy's radio was one of those benchmark designs that defined a turning point in an area of pursuit (namely portable amateur radio), much as the VW Beetle did in the world of small-car design. I built a version of Roy's radio about a year later, and it opened up a world of portable radio operation that has become my major pursuit within the radio hobby ever since. And, like I'd do with the old Beetle if I had one, I take this little radio out for a spin every once in a while just to make sure I remember what was so exciting about it way back when.

The W7EL transceiver was quite small for its day, but it was just the tip of the iceberg. In addition to the rig I had to carry along my pack, which included the rest of the station: an external battery pack, external keyer, Bencher paddles, headphones, a 40-meter dipole with 30 feet of RG 58 feed line along with all those pesky interconnecting cables. All this *just* to operate in the field. Portable is definitely a relative term.

Fast forward 23 years to the present. The Internet and the Adventure Radio Society have come into existence, the latter having issued its Trail Friendly Radio challenge: design, build and operate a QRP rig that is "friendly" to the hiker/backpacker, with an eye toward extreme portability and ease of use on the trail. While the electronics industry has booted us into the world of Giga-this, Mega-that and PIC microcontrollers, Wayne Burdick, N6KR, of Elecraft has been hard at work fleshing out a design for the *ultimate* trail-friendly radio.

The happy outcome of these events is the KX1, a radio that is so startlingly different from its forebears that it defines a new class of portable radio — the Fanny Pack rig.

Thankfully, the backpack is no longer a necessary means of conveyance. The KX1 weighs in at 16 ounces in its heaviest configuration, which includes the basic radio and optional 30-m board, automatic antenna tuner, keyer paddle and 6 onboard AA alkaline batteries. You can save a few ounces by upgrading to Lithium AA cells, which are recommended both because of their lighter weight and flatter discharge curve. And, if you desire more than 1½ W out, you can carry along an outboard 12-V battery pack and pump up the RF output to 4 W.

What do you get when you decide to spend your hard-earned dollars for one of these cute little black boxes? Surprisingly, the circuitry contains nothing startling or ground-breaking. If you can picture the basic circuit geometry of an Elecraft K1, Wilderness Radio Sierra, Small Wonder Labs DSW XX or similar NE602-based rig, you have the general idea. This route was taken as a facility for low power consumption and small size. Both goals are fully realized in the KX1. What really sets the KX1 apart are its packaging and the attention that went into selecting those special features especially useful to the wilderness operator.

The inclusion of three bands, 40, 20 and 30 m, with the optional KXB30 band module, allows operation on the two most popular contesting bands, plus casual operating at all times of the

Figure 4-3 — Elecraft's KX1 CW transceiver covers 20 and 40 meters in a shirt-pocket-size package. Add an optional 30-meter module, automatic antenna tuner and a paddle that bolts to the side of the case for a complete station. This little rig even includes a super bright LED that you can use for filling out your station log in the dark! Compare the KX1 with Cam (N6GA) Hartford's version of a Trail-Friendly Radio shown here, and you can see why Cam's new favorite Trail Radio is the KX1. **(N6GA Photo)**

day and night. The ATU option makes it possible to use just one antenna for all these bands, which is certainly a great convenience. The inclusion of a DDS VFO provides stable operation under widely varying temperatures (like those sure to be encountered on the trail), plus the bonus of being able to tune in several SWL bands when you get tired of copying CW.

Other features include a continuously adjustable IF crystal filter, 4 different tuning rates, RIT, and an LED display with adjustable brightness level. Space limitations here prevent discussing all of the features packed into this little rig, so for a comprehensive description of the KX1, check out the KX1 Data Sheet on the Elecraft Website: **www.elecraft.com**.

A few years ago, as a result of the Adventure Radio Society's Trail-Friendly Radio Challenge, several people constructed their idea of what a TFR should be. The N6GA version is pictured at **Figure 4-3**, alongside the KX1 for a little perspective. The N6GA rig featured an NN1G 40-40 transceiver, Wilderness Radio KC-1 Keyer/Counter, W6JJZ-designed Z Match, 8 cell AA battery pack and a set of Galbraith paddles. It weighed in at 2½ pounds! Compare these to the endless list of features available in the KX1, at less than half the weight, and you'll get a fairly clear picture of how far N6KR has brought us toward the goal of a completely self contained station.

***Figure 4-4* — *This photo shows the KXAT1 automatic antenna tuner board mounted above the main circuit board. Notice how the relay boxes and other components are positioned to allow assembly in a minimum of space.* (N6GA Photo)**

Building the KX1 is a breeze. There are a few surface mount parts, but these are professionally installed before the kit is shipped. All the builder has to do is spend somewhere between 12 and 15 hours installing traditional through-hole parts. In some cases, care needs to be taken to mount parts so that there is sufficient clearance inside the tight confines of the box, but the instructions are very clear about where and how. An example of this is shown in **Figure 4-4**. The main board is on the bottom, the ATU board is mounted above it, upside down. The small light-colored blocks are relays. Note how some of the relays and toroids on the two boards are interleaved between each other, making it quite a tight fit.

Elecraft does a superb job on the KX1 instruction manual. This is an Elecraft trademark: clear, concise instructions and easily readable pictorial diagrams. If you do encounter a problem, a quick detour to the Elecraft website and a chat with one of the engineers will have you quickly back on track.

One additional spin-off of Wayne's design work is the optional KXPD1 plug-in keyer paddle. This paddle plugs into the side of the rig, is secured with a thumb screw, and can be reversed to face either direction. I find that I prefer it facing to the left while operating standing up, but it works better facing to the right if I'm using the rig while sitting at a table.

My complete Fanny Pack station is shown in **Figure 4-5**. All that's needed in addition to the radio is a set of earbuds, a BNC-to-binding post adapter, and an antenna. The KXAT1 ATU is designed to tune unbalanced, random-length antennas, which are

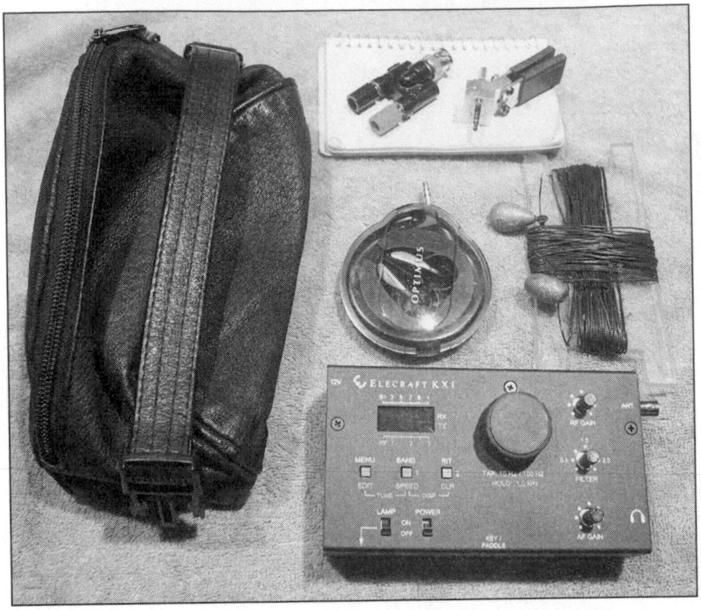

Figure 4-5 — Cam Hartford, N6GA packs an entire HF station into a small fanny pack for trailside operating. Clockwise from the left is the fanny pack, small notebook and pen with a BNC-to-binding post adapter for antenna connection and the KXPD1 plug-in keyer paddle on top, a bobbin with antenna wire and 1 ounce lead fishing sinkers, and the KX1. There is a set of "ear bud" earphones in a plastic case in the middle.
(N6GA Photo)

perfectly suited to the kind of wilderness operation the KX1 will be used for. I've settled on a 38-foot piece of number 28 wire for the antenna, with an 18-foot counterpoise. With a 1 ounce sinker tied to the end, this is a convenient length of wire for tree-tossing, and the ATU tunes it to 1.5 or less on all three bands.

Setting up the radio on the trail is slick. The most difficult step in the procedure is getting the antenna slung over the chosen branch. On a recent outing, the wire kept snagging on some pernicious little weeds and it took six tries to get the antenna up. But even with that setback, the radio was out of its pack, the antenna was up and tuned on all three bands in 8 minutes.

My best effort so far is 4 minutes from unzip to on-the-air. Because much of my trail operating is during lunch stops on day hikes with my XYL, a short set-up and take-down time maximizes the time allotted for QSOs, and minimizes grumbling about time wasted playing with radios. That's definitely a plus.

Besides being a breeze to use, the KX1 has received uniformly good signal quality reports since I've had it on the air. To top it all off, The KX1 has a built-in log book illuminating light! What more could you want in a trail-friendly fanny-pack radio?

THE ICOM IC-703 PLUS

ICOM stepped into the QRP ring recently with the introduction of their IC-703. (See the Product Review in the July 2003 issue of *QST*, page 61.) ICOM quickly followed this up by the introduction of the new IC-703 Plus, which added 6 meters to the basic low-power HF rig. What a combination: HF plus 6 meters, an internal antenna tuner and 10 watts! See **Figure 4-6**.

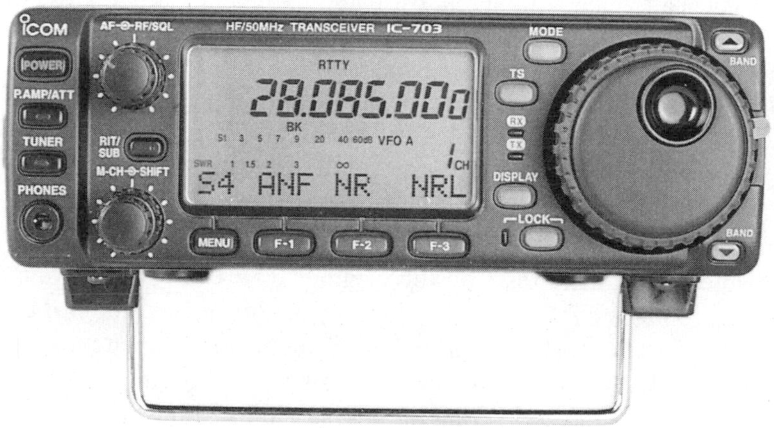

Figure 4-6 — The ICOM IC-703 HF transceiver packs a lot of features into this small package. The 703 Plus version also includes coverage on the 6-meter band.

The IC-703 Plus is a stripped-down, low-power version of their hugely successful IC-706. They dropped all VHF bands (initially) and the high power final amp and added an internal antenna tuner and came up with the IC-703. The overall look of the IC-706 is still very present on the 703. The large LCD display is easy to read at almost any angle. Controls are similarly laid out for great ergonomics.

All in all, the IC-703 Plus is a good performer. The basic IC-703 offered all the HF bands and an internal automatic antenna tuner with 10 W of RF output power. The upgraded 703 Plus adds 6 meters, which is a magic band when it comes to working with low power levels. Simple antennas and a few watts of RF are all that is needed to have some real ham radio fun whether you're running portable from the bush or at home in the comforts of your shack.

The IC-703 Plus is an easy radio to get acquainted with and an easy radio to use. The ergonomics of the controls are outstanding, which greatly speeds the learning curve associated with acclimatizing oneself to a new rig. The manual is well written and easy to understand, adding to the ease of operation.

My only 6-meter antenna is a Diamond D-130J discone that I use as the main scanner antenna at K7SZ. Using the 703 Plus and the discone I could easily bring up several of the 6-meter repeaters in and around my area. After a few QSOs using FM, I decided to check out the low end of the 50 MHz band. Using the internal antenna tuner, I was able to resonate the discone antenna system at the bottom end of 6 meters.

As usual, when I'm on the low end of 6 meters, no one else is! Granted the discone cannot be considered an adequate antenna with which to work weak signals on VHF. A set of stacked "halo" antennas would be better but a rotatable multi-element yagi antenna would be best. Since I'm fresh out of those, what the heck, let's check out the HF bands!

The IC-703 Plus is a great little HF rig. While CW keying is definitely *not* Ten-Tec QSK from the days of old, it does okay on semi-break in. Initial tests on the original IC-703 indicated some keying problems but these had been fixed on this newer model. The receiver is quite capable on 40 meters at night (my ultimate test of a receiver). I've had a lot of fun with this rig. While the 703 Plus doesn't offer all the bands that the Yaesu FT-817 does, it is still a valid choice for the QRPer who wants a portable rig and may not be interested in VHF + coverage.

THE OAK HILLS RESEARCH OHR-100

In the rush to embrace the "digital world," today's QRPers sometimes overlook a selection of analog radios that often out-perform their flashier digitized counterparts. One such product line is the Oak Hills Research analog transceivers. These are all "purists" QRP rigs, set up to work exclusively CW. Narrow IF pass bands, low-noise receivers, smooth break-in (QSK) T/R switching and ease of construction and maintenance are the big drawing points for this fine series of rigs.

Oak Hills Research was the brainchild of the late Doug DeMaw, W1FB, who started the company in 1987. Doug produced some well-designed QRP kits and accessories for a few years. He then sold Oak Hill Research to Dick Witzke, K8KEL, around 1990. Dick eventually sold the OHR product line, in 1999, to Marshall Emm, N1FN, owner of Milestone Technologies and Morse Express.

Figure 4-7 — This is the OHR 100 monoband transceiver kit. Note the small size (the AA cell is included for size comparison only) and overall clean look. **(K5FO Photo)**

The OHR-100A is a single-band (in my case 30 m) CW transceiver capable of 4 to 5 W output over a 70-kHz tuning range. **Figure 4-7** shows the insides of this little radio. This kit is offered in 40, 30, 20 and 15 m versions. The 100A features a very quite receiver section, which is definitely needed for separating the weak signals from the band noise. The 100A has a built-in RIT that covers ±1 kHz of receiver tuning range. The superhet receiver features a 9-MHz IF and a Local Oscillator (LO) running at 14.1 MHz. This places any mixing products well outside the receiver passband. The VFO is very tame and runs at 5 MHz. The receiver incorporates a 4-pole Cohn crystal filter; the crystals are hand-matched and the bandwidth is continuously variable from 1200 to 400 Hz. There is no RF amplifier ahead of the SA-602 first receiver mixer. The band pass filter feeding this mixer serves to reject unwanted signals. The OHR-100A also incorporates a very smooth QSK (full break-in keying) T/R keying circuit and has a nice-sounding sine-wave side tone. RF output power is adjustable from zero to full output via an output control on the rear panel. The SA-602 transmit mixer's output is filtered and fed into a 2N5179 buffer amplifier. From there it heads into the 2N3866 driver and finally into the 2SC2078 power amplifier. Front panel controls including RF and AF gain, RIT and IF bandwidth are nicely spaced for those of us with big fingers. The main tuning dial is easy to read. The tuning is a bit fast for my personal taste, but OHR offers an optional 10-turn tuning potentiometer as a replacement for the original tuning pot.

My overall impression of this kit is a very positive one. The large PC board offers plenty of room to work. Silk-screening is very crisp and detailed. The directions are reminiscent of "Heathkit style", very concise and easy to follow. The manual/documentation is stapled together which I think is a good thing, as this lets the builder pull the pages apart for easy access on the bench. Once completed, the documentation can be bound or punched and placed in a binder. The case is roomy, offering lots of space for modifications and additions (more on this later). In all, I am well pleased with the ORH-100A kit. Total building time from start to finish was about 18 hours.

During alignment I noticed right from the start was that set produced nearly 4 watts at the high end of the dial and less than one watt at the bottom end. It seemed like no matter how C_{120} and C_{122} were adjusted, the transmitter output was rolling off the lower edge of the low pass filter pass band, causing attenuation in the lower frequencies. A call to Marshall at Oak Hills, confirmed my suspicions. Marshall advised me to reposition the turns on the inductors L_{104} and L_{105} on the output of the NE-602 mixer (U-100) going into the base of Q103, a 2N5179. This is usually enough to get the proper power output. I tried this, but still could not obtain maximum output (4 to 5 W) from the low to high end of the tuning range of the VFO. Then I decided to lift L_{104} and L_{105} from the PC board and remove all 21 turns and rewind the coils using 23 turns. My thinking was that I would be able to remove one turn if needed. With 23 turns on both cores, I obtained 5 W output across the entire tuning range of the VFO. The added inductance widened the passband enough to allow the proper output. A check of the transmitter on a spectrum analyzer showed a nice clean output with no nasty spurs or harmonics.

On-the-air tests with this radio revealed a *very* quiet receiver. Keeping the IF filter at about 700 Hz bandwidth allowed me to hear stations slightly off frequency while concentrating on the station that I was in communications with. All too often we QRPers tend to cinch the IF bandwidth way down, really narrow, in hopes of attenuating all the signals to each side of our target station. When the IF is this tight we can't hear stations that are not quite zero-beat with our radio, thereby missing out on possible contacts. I prefer to leave the

IF opened up a bit (600 to 700 Hz) and let my brain become the CW filter, which allows me to hear off-frequency stations. This technique is especially helpful in QRP contests where congestion around the QRP calling frequency is very dense and an extremely narrow IF filter would be more of a hindrance than a help.

The transmitter is very clean and the QSK circuitry works great. No thumping as the rig is keyed at speeds in excess of 20 wpm. Tuning is a bit sharp and could benefit from the optional 10-turn pot offered by OHR. While dial markings are approximate, one worthwhile modification would be the inclusion of OHR DD-1 digital dial to provide an accurate frequency readout. The DD-1 is an external digital dial and stand-alone frequency counter in a matching OHR enclosure, featuring bright red LED digits.

The OHR-100A is a superbly capable kit radio that plays very well. The extremely quite receiver is a wonderful departure from its noisier digital counterparts. One thing I *really* like about it is its size. Over the last six or seven years we have been treated to a number of very small, compact, highly portable rigs that seem to defy easy modification or internal add-ons. Thankfully, Oak Hills Research has not succumbed to this packaging trend. You can actually work on this radio! Other OHR products include the OHR-500 five-band transceiver kit, the DD-1 digital dial, the WM-2 QRP wattmeter and the OHR dummy load. Oak Hills Research: "In the business of providing serious gear for the serious CW operator."

THE SGC 2020

While not a true QRP rig in the purist sense, the SGC 2020 HF transceiver is regarded by some as a serious contender. SGC (**www.sgcworld.com**) has had this transceiver in their product line for a few years. See **Figure 4-8**. When initially reviewed in the

Figure 4-8 — The SGC-2020 transceiver was designed by Bruce Franklin, KG7CR, formerly of Index Labs. While not simply a new and improved QRP-Plus, the 2020 is considered by many to be the second generation of that radio. It has a general coverage receiver and operates on all 9 HF ham bands. There is a 7-pole 2.7 kHz crystal filter and a switched-capacitance audio filter (SCAF) with selectable band-widths from 100 Hz to 2.7 kHz in 100 Hz steps.

October 1998 issue of *QST*, it was felt that the receiver was in need of some improvement. The newest 2020 not only has an improved receiver but the added option of audio digital signal processing to boot.

When you look at the 2020 for the first time, you are immediately struck by its unique "military styling." This rig looks like it should be equally at home in the shack, the jungle or the desert! Two words come immediately to mind: *rugged* and *bulletproof.*

The SGC 2020 is a 20-W HF transceiver capable of running CW or SSB modes on all amateur HF bands 160 through 10 m. The compact, well-built rig weighs in at 4.5 pounds and measures 2¾ × 6 × 7 inches (H × W × D). The adjustable-height front legs lift the front panel to a convenient viewing angle. The front panel features a multi-function LCD digital frequency read out and LED bar graph meter. Controls are well laid out and clearly marked. However, because of the unique way this radio operates, you'll be spending some time with the user's manual to become acclimatized to the menu selection procedures.

Power output is adjustable from 1 to 20 W using the menu buttons and the main tuning dial. Reports on my transmit audio from other stations I've contacted were almost universally considered "communications quality," meaning that the audio is very readable but definitely not "hi-fi." Like I've said before, it's about communications not high fidelity. CW keying at speeds around 25 wpm sounds good and, although the SGC 2020 is not a full break-in (QSK) transceiver, it doesn't have any bad habits like unwanted "thumps" in the headphones when keying. The T/R time is non-adjustable, so the relay chatters a little, which might be annoying to some. The built-in iambic keyer also features a speed display on the front panel. In addition, the SGC 2020 features built in SWR monitoring circuitry. These features reduce the amount of extra boxes you have to lug along when operating from the bush.

Receive audio can be tailored using the on-board adjustable SCAF (switched capacitance audio filter) and/or the ADSP option board, when installed. This ADSP board definitely augments the IF filters in the rig and can be a life saver when the bands are crowded or propagation is flakey. The ADSP board deserves special mention because it allows the user to custom tailor receiver performance to individual operating tastes. The ADSP has an automatically tunable –57 dB noise-suppression notch filter along with fully automatic spectral-noise suppression of –18 dB. The SCAF filter is adjustable from 2.7 kHz down to 100 Hz without any ringing, which can make a real difference under heavy QRM.

The SGC 2020 receiver is capable of receiving CW and SSB and will also receive short wave AM broadcasts using either sideband. The optional ADSP board can rid the incoming AM signal of annoying heterodynes, and provides easy copy of short wave stations.

As with any radio, there are faults and the 2020 is no exception. About my biggest complaint is the fact that to do anything with the menu requires the use of both hands. Menu options require simultaneous button pushing along with rotating the main tuning dial, which can be a real headache at times. The 2020 is a little hungry when it comes to current, too. Typical receive/standby current is slightly over 400 mA. Running at 13.8 V from a standard power supply or deep cycle battery, the SGC 2020 gobbles down a healthy 5 A at full power output of 20 W. Dialing back the RF output to 5 W drops the transmit current to a more conservative 1.5 A. Additionally, band changes on the 2020 can be a bit confusing at first since you must use the MEMORY function to recall a frequency back into the VFO, where it can be tuned and then restored. A bit awkward but once you are accustomed to performing the operations, band changing is relatively painless.

My final thoughts: if you are planning a DXpedition to Iraq or Afghanistan and plan on being shot at or might get sand into everything then the SGC 2020 will be right at home. Seriously, the 2020 is a "high speed, no drag" paramilitary style radio that should lend itself well to QRP operations whether safe in the shack or out in the bush.

THE SMALL WONDER LABS ROCKMITE

Dave Benson, K1SWL, of Small Wonder Labs (**www.smallwonderlabs.com**) is both a gifted engineer and a very practical man with an excellent sense of timing. Dave loves the QRP hobby and consistently provides us with innovative low-power rigs to build and use. His early success with the 40-40 transceiver kit (November 1994 *QST*) spurred him to greater heights. A succession of very interesting kit rigs poured forth from his Connecticut lab. Among them were the SW 40+, the DSW series of direct digital synthesis rigs, and several variants of PSK31 transceivers

Figure 4-9 — The Small Wonder Labs' RockMite 40-meter transceiver is a great one-evening assembly project. Packed into a MityBox aluminum enclosure (which costs almost as much as the radio kit) it makes a rugged little rig. **(Dave Benson, K1SWL Photo)**

including the amazing New Jersey QRP Club's 80 meter Warbler (*QST,* March 2001, page 37).

The DSW-II series of single-band CW transceivers use a PIC microcontroller and direct digital syntheses IC. The complete kit costs $150, and comes with all components, connectors, knobs and other parts, including an enclosure. The kit goes together in about 6 hours. The DSW-II rigs are available in 40 and 20-meter versions, with plans for 80, 30 and possibly 17-m rigs in the future.

During the summer of 2002 I had a conversation with my good friend and fellow QRPer, Chip Morgan, N3IW, who told me about the RockMite 40 meter transceiver kit from Small Wonder Labs. See **Figure 4-9**. This little "one evening kit" has a whopping $25 price tag. ***What a deal!*** Presently the RockMite is offered in 40 and 20 meter versions.

The RockMite features a direct-conversion receiver based on a Gilbert Cell mixer/oscillator (SA-612). The transmitter will deliver about 500 mW into a 50 O load. The transmit frequency offset can be toggled to provide limited frequency agility. A push of the button reverses the receive and transmit frequencies, with a total deviation of about 700 Hz.

Current requirements are very low; about 25 mA on receive and 200 mA key down on transmit using a 13.8 V dc supply. That is a very manageable power budget for operating portable from the bush using a ten-cell AA alkaline battery pack

I ordered my RockMite and the kit arrived in a couple of days, whereupon I promptly started construction. The instruction manual is a bit Spartan, but if you have several kit projects under your belt, you can safely tackle the RockMite with confidence. Most of the diodes and resistors are vertically mounted to save PC-board space. Two of the three ICs are socketed. U_1, the SA-612, is the only surface-mount component in the kit. It is easily installed, so don't be intimidated, just be careful with your soldering iron. The board is very small (2.375 × 1.875 inches) with above average component density. You must use care when soldering and the usual disclaimers about being sure that the proper components are in the right holes apply.

Total build time (less case installation) was around 3 hours. Since speed is the enemy of every homebrewer, it's not how fast you build the rig, but how fast you can get it to work, once it is built! Call me crazy, but I like to have the rig work the first time power is applied, so I take my time in the construction phase!

My 40-meter RockMite worked from the get-go and I had the little rig on the air quickly, using my 40-meter Extended Double Zepp antenna. The first thing I noticed was that the volume of the receiver was not exactly ear-shattering. Selectivity was a bit broad, too, but that was understandable. In order to test and measure RF output, you need to get the RockMite to transmit a steady carrier. Pressing one of the keyer paddles on power-up bypasses the normal keyer function and places the RockMite into the manual key mode. Closing the paddle will generate a continuous key-down condition for testing.

Default keyer speed is 16 wpm. By pressing the pushbutton switch longer than 250 ms the Morse letter "S" is heard, placing the keyer into the "speed select mode". Pressing or tapping the dot paddle increases the keyer speed (max of 40 wpm) while pressing or tapping the dash paddle reduces keyer speed (min of 5 wpm). Once in the speed-select mode, if no paddle input is detected in 1.5 seconds, the keyer emits a low tone and reverts to standard keyer operation.

With the stock 2N2222A PA transistor, RF output is slightly under ½ W. There was a lot of discussion on the RockMite Internet reflector (**rockmite@yahoogroups.com**) regarding replacing the PA with some other type of transistor. I had a spare 2SC799 in the junque box, so out came the 2222A and in went the 799. Man, what a difference! My output power jumped up to just under 1 W output (990 mW, actually)! Of course, the transmit current more than doubled, but hey, less than ½ W to almost 1 W — that's a 3-dB increase. At *these* power levels a gain of 3dB can definitely make a difference. Listening to the output on an air-local via my Drake 2B, the RockMite sounded very clean with no chirp or clicks.

Saying that the frequency agility of the RockMite is a "bit limited" is akin to stating that the sinking of the Titanic was a "boating accident." The RockMite is designed to operate on or about 7040 kHz, period. Momentarily pressing the pushbutton switch will move the frequency approximately 700 Hz by reversing the transmitter offset, but 7040 is pretty much where you are going to operate. The direct-conversion receiver selectivity is wide enough to hear people calling off frequency with no problem. Actually 7040 kHz is a good choice since it is the primary QRP calling frequency on 40 meters. The 20 meter RockMite is centered on the QRP watering hole of 14060 kHz.

The first QSO with my new RockMite netted Ohio. Not exactly rip roaring DX, but it did prove that this cute little rig was capable of making contacts. Subsequent Qs stimulated my interest in milliwatt QRP. For those of you who've never tried milliwatting, you really don't know what you're missing.

The RockMite is a unique radio that begs for a unique enclosure. Doug Hauff, KE6RIE, of American Morse Equipment (**www.americanmorse.com**), offers such an enclosure, specifically engineered and machined for the RockMite. Doug's creation, the MityBox, is an aluminum T-6061 CNC hog out that has been anodized a striking blue color. This thing is *beautiful*! The MityBox is $23 including postage.

The RockMite PC board fits snugly inside the MityBox. Three holes on each end of the box accommodate the controls and connections to the transceiver. On the front panel there are holes for the AF gain control, VFO switch, and headphones. The back panel includes holes for the key/paddles, a BNC antenna connector and the dc power connector. The MityBox has a mating cover that makes the enclosure extremely robust. Once buttoned up, your RockMite is ready for the bush. Add a set of walkman-style headphones (or ear buds), a paddle or key, antenna and small battery pack (I use a 10-AA cell pack with resetable fuse) and you are ready for some fun!

At times it can be handy to reduce the AF gain of the RockMite, especially when you're getting hammered by a nearby QRO station. By replacing R_5, a 1-MΩ fixed $^1/_8$-W resistor, on the circuit board with a 1-MΩ linear pot (Mouser pn: 31JN601) you can control the audio gain of the receiver.

The RockMite sidetone is a bit raspy, raw and excessively loud. Varying the value of C_8, a 1-μF monolithic cap (I used a 0.47-μF tantalum), will alter the sidetone level. Adding one or more RC networks between U_3 pin 5 (the PIC12C508A) and C_8 will soften this tone. The manual recommends a 10-Ω/10-μF combination as a good starting point and that worked well in my unit.

A diode in the positive supply line will prevent reverse polarity connection problems. Any silicon diode like a 1N4001 will work. Be sure that the cathode (the banded end of the diode) goes toward the RockMite board.

The RockMite can operate from an 11 to 14 V dc source using a battery or regulated power supply. If you desire to operate the RockMite from a 9-V transistor-radio battery, change R_1 to 1 kΩ and R_9 to 470 Ω.

For field outings, camping/business trips or when you find yourself on the road, the RockMite offers a complete HF station in a tiny package. While the frequency range is very limited, you can still have a lot of fun with this radio.

THE WILDERNESS RADIO SST

In 1997, Team Wilderness Radio shrank the extremely popular NorCal-40A to produce the Simple Superhet Transceiver or SST. Designed by Wayne Burdick, N6KR, as the "ultimate" backpacking rig, this little mono-bander is just the ticket for those QRPers who desire an extremely compact, low-power radio for use on the trail. See **Figure 4-10**.

Figure 4-10 — This shot of the Wilderness Radio SST's interior shows just how small a transceiver can be and still yield good performance characteristics. Of all the single band rigs currently on the market, the SST is in a class by itself. If you want a rig to design a Trail Friendly Radio around, the SST is it. **(K7SZ Photo)**

The SST comes in three flavors: 40, 30 and 20 meters. It features a three-pole crystal IF filter and VXO tuning. The tuning range is about 6 or 7 kHz on 40 meters, 10 or 11 kHz on 30 meters, and about 15 kHz on 20 meters, depending upon which varactor diode is used in the kit. There are modifications available to expand this tuning range.

There is plenty of audio in this rig to drive a pair of stereo headphones. The SST boasts an RF gain control, front-panel volume control, AGC circuitry, signal-indicator LED, and only 16 mA of current drain on receive! Power output is variable from below 300 mW to about 2 W on most models. There are also several published modifications that detail how to get slightly more power output from the SST. Basically, this consists of dropping the value of R_{10} on pin 2 of the Buffer/Driver U_5 to 150 or 120 ohms and replacing the RF amp with an MRF-237.

In my on-the-air tests, the SST receiver proved to be a darned good performer, all things considered. We are working with a very simple superhet design with a nominal parts count. While there *are* design tradeoffs with the SST, it is operator skill that ultimately prevails. The bottom line on receiver performance: don't be afraid to push the SST to the limits of its envelope. It's every bit as good as most operators will ever be!

The SST is normally powered from a dc supply between 10 and 16 V or you can use a 9-V lithium battery with a small change in circuitry, as outlined in the manual. The lithium battery can then be placed inside the SST case, producing a very compact, portable CW station.

Although the SST is at home on the trail, it has also picked up a unique, almost cultish following among QRPers who use it in the ham shack. The idea of employing a simple rig like the SST to pursue ham radio has a certain mystical, almost romantic appeal to many QRPers. After all, didn't our ham radio forefathers accomplish seemingly impossible feats of long-haul communications using rudimentary, almost primitive equipment? With the SST that thrill is back! Just ask Ade Weiss, WØRSP, who worked DXCC on 30 meters using a Wilderness SST for one third of the contacts! This is one fun radio!

Currently, the K7SZ shack is home to 40, 30 and 20-meter versions of this tiny rig. I use them for portable work as well as some genuine fun using my 40-meter Extended Double Zepp and TH7 yagi. The performance of the SST, when using a really good antenna, is amazing. All of my SSTs have the power output dialed back to 1.5 W, which is more than adequate for general CW contacts and can, upon occasion, bust a pile-up.

The SST positively screams to be modified. Not that the rigs aren't useable in the stock configuration. However, a few simple circuit changes will yield a much more "creature friendly" radio. My SST-30 is one of the original kits first produced in 1997, while my SST-20 and 40 are from the latest production run. There were some circuit changes between these two production models.

One of the first modifications I performed was to increase the IF bandwidth by replacing the four capacitors in the crystal filter. While I do not really like the stock filter, after performing the suggested factory mod to widen the filter bandwith, I noticed that the overall passband was widened somewhat, but the center frequency was shifted downward considerably. After additional experimentation, I decided to go back to the original design.

Mod number two: adding a 1-µF electrolytic capacitor to pin 7 of the LM386 audio amplifier to reduce instability under large-signal loads. This was done on the SST-30 *only*, as the later production rigs have a 2.2-µf electrolytic cap included on the board.

The SST-20 was a very different story. Even with the 2.2-µf cap from pin 7 to ground, the receiver would break into oscillation on loud signals. Out of desperation, and after a lot of troubleshooting, I replaced the LM386 audio amp IC and increased the bypass cap to a 10-µf tantalum before I was able to tame this critter.

The SST has an input for a straight key but not for paddles, so I added a $^1/_8$ inch stereo jack on the rear panel for the paddle input and mounted a K1EL K-9 memory keyer inside the SST. This nifty little PiC keyer is on a small circuit board that tucks nicely into almost any rig needing an internal memory keyer.

I definitely wanted more tuning range, especially on the SST-20. I added an SPST toggle switch to the front panel, to switch between the two varactor diodes and provide an expanded tuning range. In addition to using both varactors, I also added a second 18-MHz crystal in parallel with the original 18-MHz VXO crystal, to further increase the tuning range. This mod is not without some experimentation. Adding the second crystal can easily result in a tuning range in excess of 50 kHz! This makes tuning very difficult.

Reducing the modified tuning range requires juggling the value of RFC_3 in the VFO. I reduced the value of RFC_3 from 5.6 µH to about 4.7 µH. The total VXO swing, switching between the two varactor diodes, is now about 28 kHz. This provides coverage from 14.039 to 14.066 MHz with only 3 kHz of overlap. Not too bad for a crystal-controlled rig!

Soldering an 18-kΩ resistor across the 10-kΩ main tuning pot (from the wiper to the high-voltage side of the control) does a lot to linearize the tuning. This results in signals that are spread out over the rotation of the tuning control and not bunched up at one end or the other.

Neither varactor diode would give me the tuning range I wanted (10.105 to 10.118 MHz) on the SST-30. I substituted another 14.318-MHz crystal (20-pf load capacitance), for the standard series crystal furnished with the kit, but still had the same tuning range. I didn't want to use a switch to select between the two diodes, so I paralleled a second crystal across the existing one and ended with a range of 10.097 to 10.120 MHz, using the MV-209 varactor. This was ideal for what I needed so I left well enough alone. I had to reduce RFC_3 from 12 μH to 9.6 μH by using a 6.8 μH and a 2.7 μH inductor in series. I did not need a linearizing resistor on the SST-30, as the rig produced a tuning range of about 11 kHz on each side of the control.

I have used both rigs to make lots of DX contacts and other QRP QSOs. There is really nothing quite like the SST. It's a perfect blend of simple, analog technology and gutsy design. Its diminutive size belies its capabilities. This little rig's got a lot of heart!

In conclusion, the Wilderness Radio SST kit offers both the neophyte and experienced home-builder a great project that combines innovative engineering with an RF platform that begs to be modified. Can this Minimalist Rig be considered a "real radio?" I definitely think so, and so do many others who've used the SST for any length of time. Once you build and modify an SST, you will have the satisfaction of using a rig that is tailored precisely to your particular operating style. The price is definitely right: $89!

THE WILDERNESS RADIO NORCAL-40A

The Wilderness Radio NorCal-40A transceiver kit has been on the QRP scene for about 8 years. Since 1995 the NC-40A has been a mainstay of the Wilderness line. Well over one thousand of these kits have been purchased by QRPers all over the world. In addition, undergraduate students of the electrical engineering program at Cal

Tech build the NC-40A as a project each year. (*The Electronics of Radio*, by David Rutledge, KN6EK is the text for this course. The book is available from ARRL.) To say that this kit has some mileage on it is an understatement. **Figure 4-11** shows a NorCal-40A.

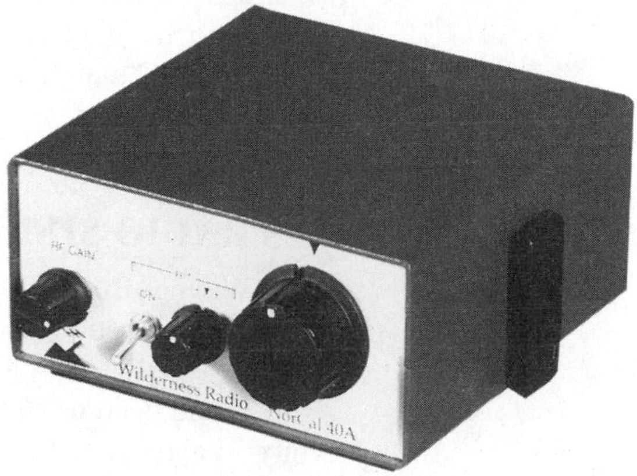

***Figure 4-11 — The NorCal 40A by Wilderness Radio is a wonderful little single band QRP radio. Performance is amazing, considering the simplicity of design and parts count. My NC-40A accompanied me on an 8000 mile trip around the US and Canada in 1996. I operated QRP mobile on 40 m from the car and made over 30 contacts using only 3 W and a mobile whip antenna.* (K7SZ Photo)**

As a first-time builder's QRP kit, the NC-40A shines. The small circuit board is well laid out and silk screened for easy assembly. There are only a handful of toroidal inductors to wind.

First of all, the NC-40A is a very well mannered rig. The stock transceiver can give the QRPer many hours of pleasure and excitement, working the world on 40 meters. If you want to customize this little sleeper, however, you can certainly end up with a "street rod" that will yield amazing performance at reasonable cost.

The stock NC-40A's nominal RF output is around 1.5 to 2 W with the 2SC799 PA transistor furnished with the kit. If you want a bit more power out of this rig, I suggest substituting a Motorola MRF-237 transistor for the PA. This transistor is becoming scarce and a little expensive, but I feel that this mod is well worth the effort and expense. Power output with the new PA climbs to around 3 W. Of course, the transmit power budget is also increased, but that's the price you pay for about 3dB of gain.

Under no circumstances should you "play" with the LC values of the output filter network. There were several mods described in 1996 dealing with substituting capacitor values in the output filter to increase RF output. This is very dangerous. Spurious transmitter products cannot be measured unless you have access to a spectrum analyzer. An increase in spurious emissions will look just like an increase in RF output power to a simple wattmeter. Therefore, unless you can see the output with a spectrum analyzer, it is best to leave the calculated stock LC values of the output filter alone.

The Wilderness Radio NC-40A is one heck of a deal for less than $130. This cost includes all the parts plus a case, knobs, jacks and controls. This rig has a solid, field-proven track record. It's also the rig "that started it all." As with all Wilderness products the support is first rate. You can buy and build this rig with confidence.

THE WILDERNESS RADIO SIERRA

Wayne Burdick, N6KR, (the "radio" half of Wilderness Radio) designed the NorCal Sierra, a new-generation analog multi-band CW QRP kit built with portability in mind. By mid 1996 the Sierra had become *the* low power rig of choice for the QRP fraternity. Wayne stated that his primary goal was to design a kit radio with the backpacker/hiker in mind. The Sierra was smaller, lighter, and more battery friendly than other QRP rigs of the time, *and* it offered multi-band capability for the backpacker/hiker.

Wayne and Bob Dyer, KD6VIO, formed Wilderness Radio as a commercial kit radio company and started offering the NC-40A and the Sierra as part of their product line. Sales went through the roof. In a very short time the Sierra had established itself as the premier QRP transceiver of the '90s. In addition to being a great portable rig, the Sierra also holds the Miles-Per-Watt record on 40 meters, set in 1994 and shattered again in 1995 by the low power team of Fran Slavinski, KA3WTF and Paul Stroud, AA4XX.

The June 1996 issue of *QST* includes a review of the Wilderness Radio Sierra. While the Sierra covers all nine ham bands, you only need to purchase the bands you are interested in, thereby customizing the radio to your particular operating habits.

The Sierra covers 150 kHz on each of the nine HF ham bands, depending upon the band module installed. The innovative idea of using band modules eliminates the need for complex band-switch wiring, which, in turn, greatly simplifies construction. The elimination of band switching components like PIN diodes and relays further serves to reduce current consumption. Speaking of current, the sierra draws about 30 mA on receive (using headphones) and up to about 450 mA on transmit, depending upon RF output.

The receiver uses Gilbert cell mixer technology in the form of NE602 (NE612) mixer/oscillators. Manufacturer's claimed specs are: minimum discernible signal (MDS): –135 dB, blocking dynamic range (BDR): 110 dB, Two-Tone DR: 88 dB. These are quite respectable performance parameters, considering the simplicity of the overall design. Selectivity is variable between 150 and 1500 Hz thanks to an innovative variable bandwidth (ABX) feature that is adjustable from the PC board or that can be remotely connected to the front panel. This provides an easy way to cinch down the IF bandwidth in crowded band conditions. The ABX controls the passband of a 4-pole Cohn crystal filter. There is an additional single-pole crystal filter after the IF amplifier circuit.

The receiver also features an RIT circuit that moves the receive frequency approximately ±2 kHz away from the transmit frequency. There is a mod, discussed in the construction manual, to increase this RIT swing to around ±10 kHz by changing R33. This mod will enable you to operate with enough frequency agility to bag most DXpeditions using split-frequency operation.

All in all, the sierra is a very well-behaved radio that offers outstanding performance in a relatively small, 2.8" (H) × 6.6" (W) × 7.0" (D) package. Weight is around two pounds. The rig will run on a dc supply anywhere from 10 to 16 V.

Power output is between 2 and 3 W depending upon the band and the supply voltage. The RF output tends to be reduced on 15 meters and higher frequencies. My current Sierra has a PA mod: I replaced the stock 2N3553 transistor with a Motorola MRF237. This boosts my power output to over 3 W up to 15 meters and a solid 2.5 W on 15 meters and higher frequencies.

The VFO, a series-tuned Colpitts oscillator, operates between 2.935 and 3.085 MHz, providing excellent linearity over the 150-kHz tuning range. An 8:1 vernier drive slows the tuning rate to an acceptable level, providing easy tuning of incoming signals. Advertised drift is less than 100 Hz after a 15-minute warm up time. In reality, I have noticed virtually no drift from a cold start on any of the three Sierras I've owned. This is an important feature for portable operation, where the rigors of the bush coupled with ambient temperature variations can wreak havoc with less-robust VFO designs.

Wilderness Radio offers several options for the Sierra that will definitely enhance operation. First is the KC2 digital display/

Figure 4-12 — The Sierra, from Wilderness Radio, is a very popular rig with great performance characteristics. Multiband operation (160, 80, 40, 30, 20, 17, 15, 12 and 10 m) is realized using band modules. On the left is the analog-meter version of the Sierra, with the digital-display version shown on the right. The Sierra was used by Fran Slavinski, KA3WTF, and Paul Stroud, AA4XX, to set the world mile-per-watt record on 40 meters!

memory keyer module. Installing this module requires a new front panel for the Sierra. Wilderness furnishes a pre-punched and silk-screened front panel that really dresses up the little rig. The KC2 features two programmable keyer memories, while the display functions as a digital frequency display, and also reads power output and S-meter indications. **Figure 4-12** shows both a plain Sierra and a Sierra with the KC2 option installed.

The Buzz Not is a novel little noise blanker that installs in the Sierra IF strip. This is a variable-width-pulse noise blanker that is quite effective in cleaning up short-duration impulse-noise bursts.

The Wilderness Radio Sierra is a great performer. The receiver is exceptionally quiet, allowing you to dig the weak ones out of the band noise. Hundreds of Sierra owners regularly use their rigs to chase DX, contest, rag chew, participate in beacon competitions and operate from the bush. This little rig is a very versatile transceiver that can fill the bill as the main station radio or a tag-along rig for camping and hiking trips.

YAESU FT-817

Seldom have I been captivated by a radio. The Vertex Standard (Yaesu) FT-817 is the exception. The FT-817 provides the QRPer with an extremely flexible HF/VHF/UHF radio platform that is at home in the shack, in the bush or on top of a mountain.

This diminutive, feature-packed rig is sooooooo much fun to operate! To get an accurate feeling of how this radio performs check out the Product Review in April 2001 *QST*, page 75, and the expanded lab reports on the 2001 *QST* CD ROM.

Briefly, the radio covers all the HF bands (160 to 10 meters) and the receiver will function as a "world-band receiver" allowing you to listen to AM SW broadcasts, military long-haul flight following SSB transmissions, and about anything else that goes on in the HF spectrum. The FT-817 offers the following modes: CW/AM/SSB and Data. This radio is so diverse in its frequency and mode coverage that it is scary! Power output is controllable from the front panel via menu selections. There are a number of layered menus involved with operating the rig, but once you have the basics down it is a fun radio to use. There are several aftermarket "cheat sheets" available for use in the bush or while mobile/portable. They take the guess work out of remembering what each menu does.

In addition to general QRP fun, this is a serious ARES/RACES emergency communications station. This rig does it all and does it well. QRP power levels should not deter an emergency communicator from buying this rig to augment his station equipment. If needed there are both HF and VHF/UHF amplifiers (ugh, there is a

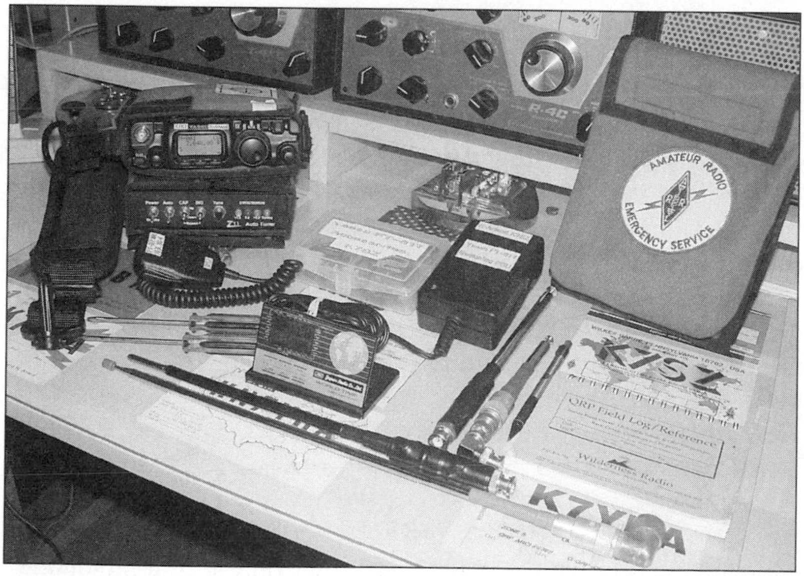

Figure 4-13 — The Yaesu FT-817 transceiver covers all bands from 160 meters to 70 centimeters. Here, the contents of the author's "travel kit" are spread out on his operating desk. (K7SZ Photo)

nasty word!) that will boost the FT-817 output into the 35 to 50 W range. Face it, if you need the extra RF in an emergency, don't be bashful or proud; use the amp.

I have taken my FT-817 on numerous business trips and family outings plus several long trips to Florida and Georgia. This radio has never let me down. I have used it portable and mobile and in all sorts of climatic conditions. See **Figure 4-13**. It is a great rig that has provided me with a lot of fun and many, many QRP QSOs and contest contacts.

CW operation is not full break-in, but is tolerable even with the chattering relay. Using semi break-in and setting the TR switching at about 250 ms, I am able to zip right along at 25 wpm with no aggravation. The inclusion of the 500-Hz Collins CW crystal filter is a must for any CW operation. This filter makes a big difference in receiver performance and ease of use in the CW mode. There are a lot of small companies that offer after-market options for the FT-817, so shop around for the best deal on filters, battery packs, charger/power supplies, etc.

I have researched ideas and modifications that might prove useful to further adapt the 817 to life on the go. What follows is just the tip of the iceberg.

The strap provided by Yaesu for the FT-817 is difficult to remove and always gets in the way when operating. I solved this problem by adapting a set of "D" rings and shoulder strap from a VCR camcorder bag. The strap can be clipped onto the "D" rings to provide a wider, much more comfortable carrying strap, which can be completely detached during on-air operations.

The standard Yaesu battery pack (FBA-28) is designed to hold eight AA cells. Unfortunately alkaline cells are quickly depleted. The optional 9.6 V 1000 mA/h NiCd pack (about $60) doesn't last as long as the alkaline cells, but can be recharged up to 1000 times or more.

One suggestion regarding the FBA-28 battery pack was to purchase 1600 mA/h NiMH "AA" cells and build up a battery pack in that fashion. Unless you modify the plug on the FBA-28 by cutting or disabling the green wire, you won't be able to charge the NiMH cells without removing them from the radio. This is extremely unhandy.

An economical solution exists at Radio Shack. Purchase their 9.6 V 1600 mA/h battery pack (RS # 23-331, $24.99), remove the plug and replace it with the plug from the FBA-28. You now have a NiMH pack that is internally rechargeable at a sizeable cost savings compared to the $60 after-market NiMH packs!

Of course, you can (and I highly recommend you do) procure a large (6 to 7 Ah or larger) gel cell as a portable power source to conserve your on-board battery pack in the FT-817. These gel cells can be found at hamfest flea markets at reasonable prices. Another great source of these batteries is your local medical equipment repair and/or burglar alarm company. Many times these folks routinely replace these gel cells on a cyclic basis. The batteries still have lots of life left in them and can be purchased at minimal cost and sometimes these folks will give the batteries away for free just to eliminate recycling fees!

For fixed station use, Jameco Electronics (**www.jameco.com**, telephone: 800-831-4242) offers a compact 12 V dc, 3.3 A (P/N: 155213, Prod #: P40A-3P2JU) computer switching power supply for only $29.95 (plus s/h). Measuring 5.5" × 2.3" × 1.5" and weighing less than 8 ounces, this diminutive supply powers the FT-817 nicely. It's small enough to fit into your "jump kit," providing you with a means to power your radio as long as ac mains are available. I haven't noticed any RFI in my trials, which included extensive operation on 10 meters! You'll have to change out the dc connector to match the FT-817 power jack, but for the price, this little PSU is a great buy.

Without a doubt the FT-817 needs help in the filter department when operating CW. The stock filter is much too wide to provide the selectivity needed for even casual CW operation. Yaesu offers their 500-Hz Collins mechanical filter. This is a seven-pole filter that really improves the CW performance of the receiver. It is a "drop-in" mod only requiring removal of the cover on the 817, and it only takes about 10 minutes including removing and replacing all the screws! A quick trip to the menu to enable the optional CW filter and you're in business! What a difference it makes! The filter can be bypassed by a simple push of a button to allow for wider bandwidth. Plan on purchasing this filter with the radio. It will save you time and frustration, especially if you work a lot of CW.

Unfortunately, the FT-817 offers only *one* optional filter slot. It would have been nice had they planned ahead and offered a second slot for the optional Collins 2.3-kHz, 10-pole mechanical filter, so both SSB and CW operation could be enhanced. As it sits now, as a new FT-817 owner, you are relegated to either the optional CW or SSB filter but not both.

Enter W4RT. The good folks at W4RT Electronics offer a mod that replaces the stock Yaesu SSB filter with an optional 2.3-kHz, 10-pole Collins mechanical filter *and* keeps the optional Collins CW filter in place. Phone receiver performance is substantially improved and SSB transmit audio reports are said to be better than those received using the stock filter. Unfortunately this mod (which is performed by the vendor) costs half as much as the FT-817 (around $300 plus s/h at the time of this writing)! I have corresponded with several QRPers who've done this mod and, price aside, they all like it...a lot! So that is something to think about when you are planning which optional extras to include on your FT-817.

The transmit audio solution might be somewhat simpler and realized at a much lower cost. Discussions with members of the FT-817 groups have yielded some insight into this SSB filter situation. Apparently, the increase in transmit audio performance is directly related to the reduction of wasteful low-frequency audio going into the transmitter. By using the Heil HC-4 mic element, you can have a similar improvement in transmit audio for about $^{1}/_{10}$ the cost!

I have used a converted HT speaker/mic since I bought my FT-817. The mic has been modified by removing the speaker and disabling the electeret mic element. I added a Heil HC-4 mic element, and the cord was terminated in an RJ-45 plug to mate with the mic connector on the 817. The results have been fantastic! I have received very complimentary audio reports from stations I've worked both on SSB and FM. The hard-hitting mid-range audio provided by the HC-4 element really punches through. Face it, you need all the help you can get when operating SSB on a QRP rig. The aggressive audio provided by the Heil HC-4 mic element really "gets the message across." Hey, it's about communications, not hi-fi!

If I were to be told that I could have only one ham rig for the rest of my life, I would unhesitatingly choose the Vertex Standard (Yaesu) FT-817 as that do-all radio. Sure, as with any radio, there are some shortcomings but the plusses far outweigh the minuses on this radio. Portability, easy of use, band and mode coverage, flexibility and rugged construction make the FT-817 a serious rig for serious QRPers.

THE MFJ 90-SERIES

Martin Jue, K5FLU, enticed Rick Littlefield, K1BQT, to design a series of single-band CW and SSB QRP rigs for MFJ Enterprises. The MFJ units come fully built and ready to "plug and play." Each rig covers a specific band: 9040 — 40 m, 9020 — 20 m, etc. There are rigs for 40, 30, 20, 17 and 15 m. Each rig covers 500 kHz of the lower end of their respective band. The receiver section is a superhet design with crystal filtering in the IF strip that features audio derived AGC and RIT. Keying is semi-break-in and not full QSK.

The MFJ 9400 series transceivers are the SSB versions of the 9000 series CW rigs. That's right, MFJ markets four single-band sideband HF rigs that are great for the car or the shack. See **Figure 4-14**. Based on Rick Littlefield's design, these small radios offer the sideband operator a chance to work QRP while traveling or camping. These radios also feature a superhet receiver with audio-derived AGC, crystal filtering and an optional CW module that will turn them into a CW/SSB rig! Current production 9400 series rigs cover 75/80, 40, 20 and 10 m.

These little single banders are great for camping, business trips and general traveling around. A small gell-cell battery, key

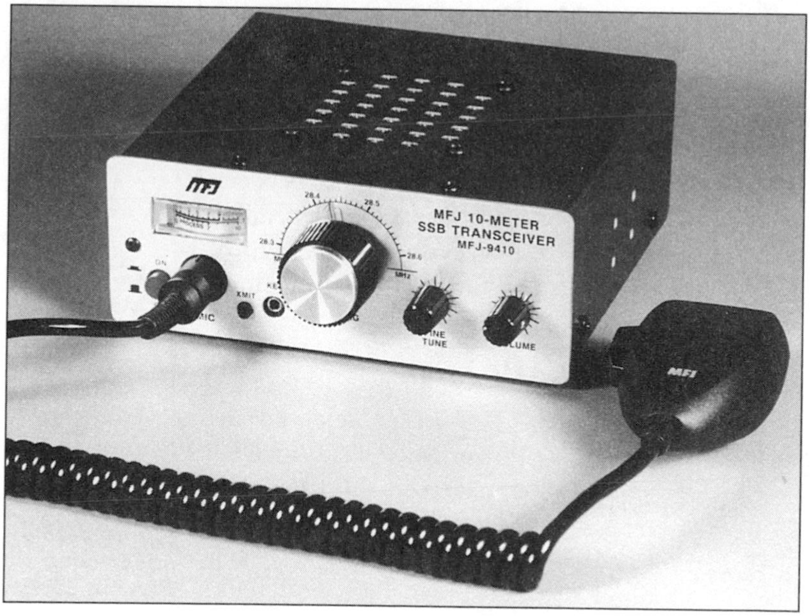

Figure 4-14 — The MFJ 9410 is this company's latest single band QRP SSB transceiver. Controls are ultra simple, making this a great radio to have along during camping and portable operations. Output is about 8 W, but can be throttled back to provide 5 W output to meet true QRP requirements. **(K7SZ Photo)**

or keyer, headphones and wire antenna are all that is needed to "take it on the road."

For the VHF QRPer, MFJ also offers a 6 and a 2 m SSB/CW version (9406 and 9402, respectively) that put out about 5 W and are great for mountaintopping. QRP Field Day groups can use these two rigs to add VHF capability to their efforts.

CLASSIC QRP GEAR

What follows is a quick look at the rigs that have made a lasting impression on the QRP community over the years. Here you'll find some real classic rigs that still work well on today's crowded bands. One word of warning: collectors have taken over the used QRP gear market. Prices for classic QRP gear have risen dramatically over the last few years thanks to the efforts of collectors to obtain items for their "collections." Case in point: at the York (Pennsylvania) ham-fest a few years ago I picked up a complete Ten-Tec Argonaut 509 station consisting of the 509 transceiver, matching 251M power supply, model 405 linear amp (50 W), Ten-Tec ceramic microphone and crystal calibrator, including all the original manuals, for under $300! Today that station would net between $600 and $800, depending upon cosmetic and electrical condition! Caveat Emptor!

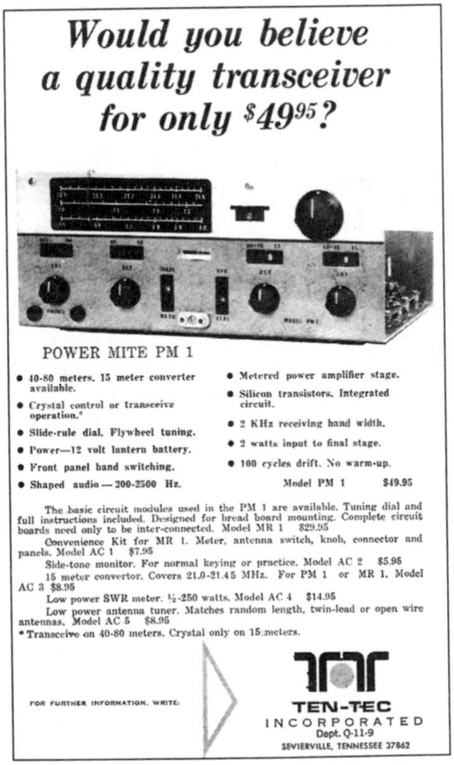

Figure 4-15 — Ten-Tec began advertising the Power Mite model PM 1 in 1969.

TEN-TEC QRP RADIOS

Ten-Tec released the Power Mite series of QRP transceivers in 1969. See **Figure 4-15** for an early ad for the PM-1. The PM-1, PM-2 and PM-3 offered a choice of band coverage along with a direct conversion receiver and about 1 to 2 W output in an attractive little cream and woodgrain box. The slide rule dial was silk screened on the front of the case. The receiver section was a single ended MOSFET configuration and was very prone to ac hummmmmm, AM breakthrough and microphonics. All these rigs bring premium prices on the used market today, when you can find someone who wants to part with one. Be prepared to pay $125 and up for a PM rig! **Figure 4-16** shows my PM-1 and PM-3A rigs.

Without a doubt, the venerable Argonaut series of rigs is synonymous with low power communications. Ten-Tec introduced the Argonaut model 505 in 1972. The 505 covered 80 to 10 m on CW and SSB with a power output of 2.5 W and featured a permeability tuned oscillator in the VFO (these specs are virtually the same on all analog Argonauts). The superhet receiver with crystal filtering was a boon to QRPers who were used to the direct conversion receivers of that era. An audio derived AGC and S/RF Power/SWR meter provided convenience, especially when working portable. There are not many 505s in circulation on the used market; therefore the prices vary widely

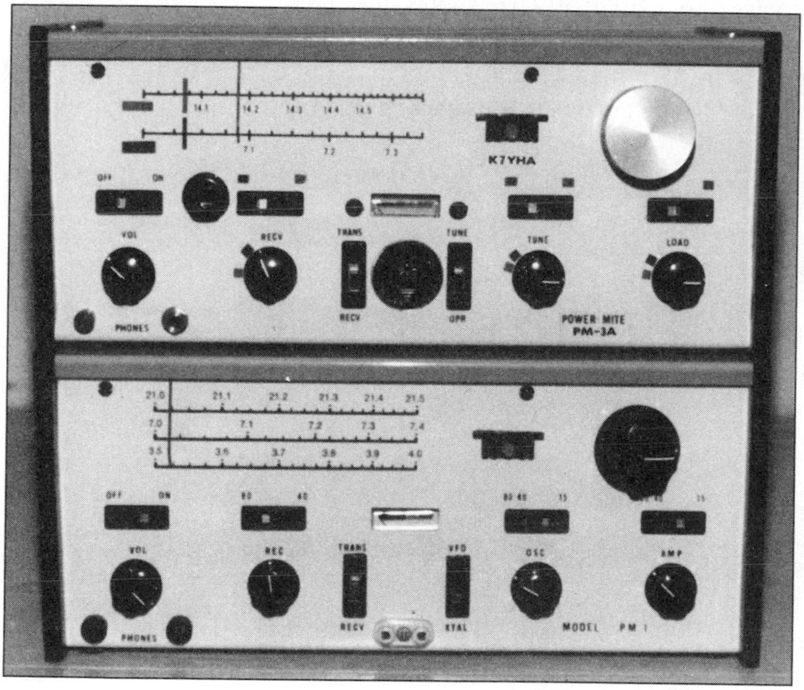

Figure 4-16 — These two Ten-Tec PM rigs (a PM-1 and a PM-3A) lived in my shack for a few years. They were Ten-Tec's introduction to the world of QRP in 1969 and into the early 1970s. I used the PM-3A when stationed in the UK to work about 50 of the G-QRP Club members on 80 and 40 m. Forty, espe-cially, was a real treat owing to all the short-wave broad-cast stations that totally swamped the receiver on occasion. **(K7SZ Photo)**

Figure 4-17 — The Ten-Tec Argonaut 509 is one of the most popular QRP radios ever produced. While their performance is not dazzling, they certainy command a high price on the used market. This rig has the legendary Ten-Tec full-break-in keying, about 2.5 W output on 80, 40, 20, 15 and 10 m, and a mediocre receiver. Prices are high when you can find a used one. **(K7SZ Photo)**

depending on condition. With the recent nostalgia craze sweeping the QRP world, be prepared to pay premium prices for these semi-rare rigs.

The 505 was followed by the model 509 in 1975. This is the most common of the Argonauts with several thousand being manufactured. The 509 featured several upgrades including a broadbanded transmitter section. Prices start around $300 for a used 509 but can go rocketing upward depending upon accessories and condition. See **Figure 4-17**.

The model 515 debuted in 1980 and was the last of the analog Argonauts. Ten-Tec made significant improvements in this radio, which included splitting up the 10 m band into four 500 kHz segments along with a much quieter receiver. The 515 was a radical departure from previous Argonauts in that the black and gold color scheme was quite striking in contrast to the beige and wood-grain cabinets of its predecessors. **Figure 4-18** shows this new cabinet design. There were a limited number of 515s produced between 1980 and 1983, so prices are astronomical and completely out of line with the performance specifications of this radio. Average prices on this rig start at $450 and go up!

We'll round out our Ten-Tec coverage with three more of their

Figure 4-18 — The Ten-Tec Argonaut 515 is the last of the analog Argonaut series. Notice the styling difference between the 515 and the 509. One thing Ten-Tec did right was to expand the 10 m band into four 500 kHz segments. Improved receiver designand a great look make this the QRP rig most coveted by Ten-Tec collectors. Only 800+ were ever made, so the prices on these rigs are extremely high. **(K7SZ Photo)**

offerings. First is the Argosy-I/II (models 525 and 525D). The Argosy series is one of the best kept secrets of QRP and is considered by most Ten-Tec aficionados as the natural evolution of the Argonaut. First introduced in the early 1980s, they were never very popular to the general ham radio market, but they have a fiercely loyal following by their owners. The 525 was an analog model and the 525D was the digital readout model. See **Figure 4-19**. There were at least four revisions of the internal circuitry on this radio series, but they all play pretty well. Several modifications exist to improve receiver performance and, provided you are a competent technician, are well worth the trouble to perform. The nice feature of this radio is the ability to go from 50 W output down to 5 W output with the flip of a switch. Buying an Argosy gives the QRPer the best of both worlds. Band coverage includes 80, 40, 30, 20, 15 and 10 m. Options include a noise blanker (quite ineffective), active audio filter (extremely effective) and a choice of several IF crystal filters. One very worthwhile addition is a T-Kit speech processor inside the Argosy, which really makes a difference in how this radio performs in the SSB mode. Prices start around $350 for the analog model, slightly higher for the 525D (digital) model with matching power supply.

Figure 4-19 — The Ten-Tec model 525D Argosy-II (digital version) is one of my favorite QRP rigs. This is a six band radio (80, 40, 30, 20, 15 and 10 m) that offers SSB and CW modes, and can go from QRP power levels to 50 W output at the flick of a switch. This radio is like having the best of both the QRP and QRO worlds. Keying is full-break-in and performance is quite good overall. **(K7SZ Photo)**

In the late 1980s the QRP world kept hearing about the "new" Argonaut soon to be released by Ten-Tec. After three years of factory hype and promises, the Argonaut-II, model 535 finally made its debut. Unfortunately, this over $1200 QRP radio was doomed from the start. Few people wanted to pay $240/W for a QRP rig!

The Argo-II was technically a Delta-II transceiver with the 100 W RF deck stripped off. The 535 had a lot of big rig features including microprocessor control, wide-band general coverage receiver (100 kHz to 30 MHz), multifunction digital display, direct keypad entry of frequencies, continuously variable IF bandwidth and much, much more. See **Figure 4-20**. Due to a series of problems with this radio, it received a rather bad reputation early on, from which it never recovered. Production was halted in 1995. Ten-Tec released several versions of microprocessor firmware. Used Argonaut-IIs are common on the swap nets and Internet trade lists with prices starting around $700.

The last Argonaut produced by Ten-Tec was a variation of their Scout, called the model 556. This rig was basically a Scout (model 555) sans the 50 W RF deck. The 556 was a unique radio

Figure 4-20 — The Ten-Tec Argonaut 535 (Argonaut-II) was a full-featured QRP rig that was marketed for several years. Basically a Delta-II without the 100 W RF deck, this rig had a host of bells and whistles, including fully variable IF bandwidth and computer control. Although never really popular with the QRP set, owing to its $1200 price tag, this rig can hold its own, performance wise, with any of the imports. **(K7SZ Photo)**

in that the rig was essentially a single-band radio featuring plug-in band modules. Therefore, you could order the radio with only the bands you really wanted to work. Designed primarily as a mobile/portable transceiver, the 556 was made for the ham on the go. Unfortunately, this little rig is no longer in production. Prices on the Internet reflectors start around $400 and up depending on the number of band modules and optional accessories.

One final word about Ten-Tec radios. Their full-break-in (QSK) keying is legendary. All the Argonaut rigs, along with the Argosy series, offer very smooth break-in keying. Once you get used to this feature, going back to semi-break-in is no fun at all.

Ten-Tec has directly supported the QRP fraternity since the late 1960s. They are to be applauded for their efforts. Having talked with several people in the upper echelon of Ten-Tec, I have been enlightened as to the costs of manufacturing a QRP rig versus the sales the rig will generate. In a nutshell, it's a business disaster. Basically, it costs the same to research, design and produce a QRP rig as it does a 100 W radio. With the limited market in the QRP arena, it's a "no brainer" to visualize that a QRP rig will not generate the revenue necessary to offset the costs of production (something that none of the Ten-Tec QRP rigs has ever done!). So, before you criticize Ten-Tec or any other major manufacturer for not producing a state of the art QRP rig, just remember, it doesn't make good business sense to do so and should not be taken as a reflection of how the company feels toward the low power side of the hobby.

THE HOT WATER SERIES FROM HEATHKIT

Heath Company in Benton Harbor, Michigan, marketed three Heathkit QRP rigs, starting in 1973 with the HW-7, which covered 40, 20 and 15 m and had about 1.5 W output. The HW-7 is not a useable rig on today's crowded bands, but in its day it had quite a following. The HW-7 receiver was patterned after the Ten-Tec PM series and featured a single ended MOSFET front end that suffered horribly from ac hum, AM breakthrough and poor selectivity.

In the mid-'70s Heathkit marketed the new and improved HW-8. This rig featured a completely redesigned front end (it was still a direct conversion receiver but the overall specs were much better than the abysmal HW-7) and included 80 m coverage. The HW-8 was in production the longest of all the "Hot Water" QRP rigs and had the distinction of being the most heavily modified QRP radio in the

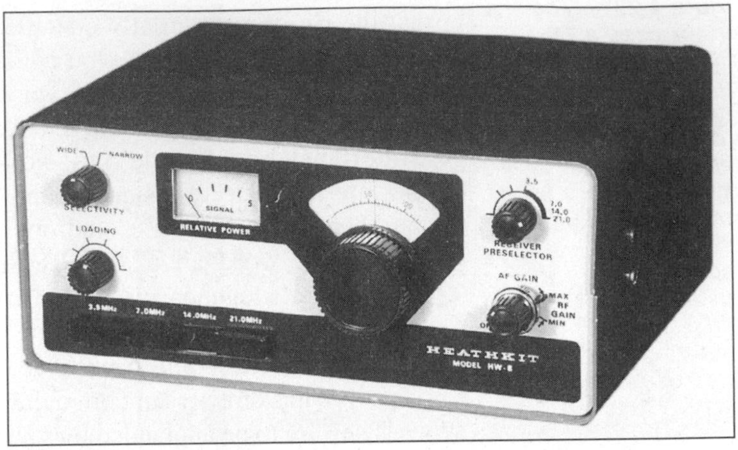

***Figure 4-21 — The Heathkit HW-8 was the most highly modified QRP rig in history. This rig has four-band coverage, a decent direct-conversion receiver and about 2 W output. The keying is relay controlled so it is definitely not a full-break-in radio. Lots of these radios find their way to ham radio flea markets, so look around and grab a piece of QRP history.* (K7SZ Photo)**

world! See **Figure 4-21**. *The Hot Water Handbook*, first produced by Fred Bonavita, W5QJM, and subsequently rewritten and revised several times by Mike Bryce, WB8VGE, is the bible on how to wring the most out of this little rig.

The HW-8 was superseded by the HW-9 in the early 1980s. The HW-9 was a radical departure from earlier Heathkit offerings and had its share of problems: namely receiver selectivity and an unstable transmitter on 10 and 15 m. Of the three, the HW-8 is the best choice and these can be found at hamfest flea markets for around $100. The HW-7 and 8 models both had direct conversion receivers. The HW-9 is the only Heathkit QRP rig to feature a superhet receiver and all nine HF bands. See **Figure 4-22**. Prices on the HW-9 range from $175 to over $300 depending on cosmetics and options.

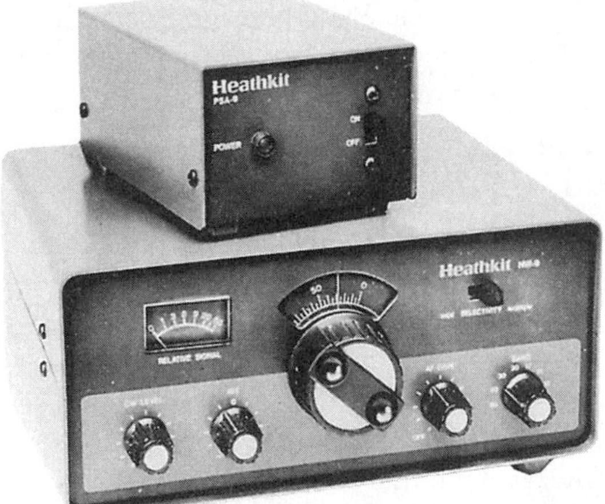

Figure 4-22 — The last of the Heathkit QRP rigs, the HW-9 covered the lower 250 kHz of the 80, 40, 20 and 15 m bands. There was also an optional accessory pack that covered the 30, 17 and 12 m bands as well as the lower 250 kHz of the 10 m band.

THE KENWOOD TS-120V/130V

The Kenwood TS-120V/130V are two choices that can be found on the used market occasionally. See **Figure 4-23**. The difference is the added 30, 17 and 12-m band coverage on the 130V, along with selectable dual crystal IF filters for both CW and SSB, a 20

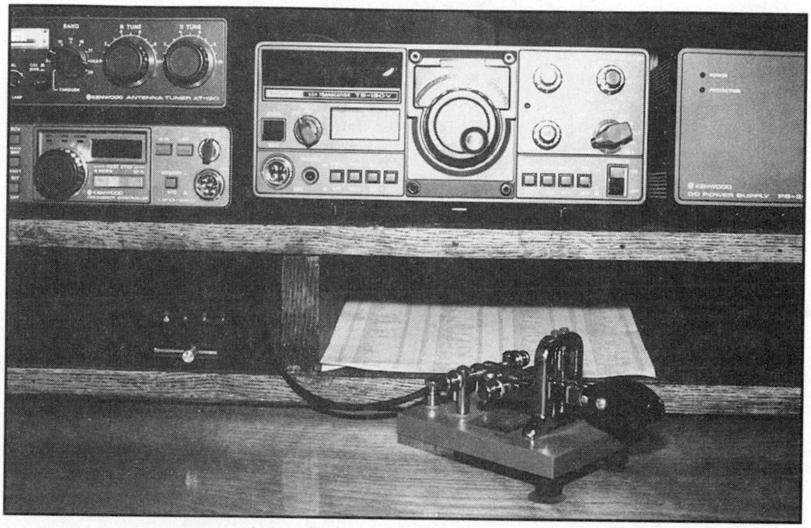

Figure 4-23 — The author used a TS-130V in his station for several years. These rigs featured a digital display and selectable dual crystal IF filters for both SSB and CW. Optional accessories included an analog and a digital VFO, antenna tuner and power supply. **(K7SZ Photo)**

dB RF attenuator and a speech processor. Both radios are 10 W versions of the 100 W TS-120/130S models. They can easily be pulled down to 5 W in both CW and SSB modes. These rigs feature digital readout and have a load of optional accessories including both an analog and a digital VFO, external speaker, power supply, antenna tuner and outboard 100-W linear amp. Prices vary considerably from East to West Coast, so do your homework before you plunk down your cash. The last TS-130V I bought cost $350 without any accessories.

QRP-PLUS

The Index Labs QRP-Plus transceiver was produced from 1994 through 1996. Bruce Franklin, KG7CR, was the brains behind this QRP rig. The QRP-Plus features a general coverage receiver, 160 through 10-m ham-band coverage, full microprocessor control of the radio, low current drain (140 mA in receive and 1.5 A in transmit), and very small physical size (5.5 × 4 × 6 inches). The large LCD digital readout is easy on the eyes and the steel case provides an extremely rugged enclosure. See **Figure 4-24**. The superhet receiver features a switched capacitance audio filter (SCAF) that works extremely well. Power output is a full 5 W, which can be adjusted down into the milliwatt regions

Figure 4-24 — The Index Labs QRP-Plus is a sharp looking rig. With full ham band coverage on both SSB and CW, this rig also includes a general coverage receiver. Fully microprocessor controlled, the QRP-Plus has a host of features including built-in SCAF filtering, multiple tuning rates and a large LCD frequency readout. Although it was only marketed for a short while, this rig gained rapid acceptance by the QRP fraternity. **(K7SZ Photo)**

for some really "hard core" QRP! Optional accessories included a matching power supply/antenna tuner. Prices start around $350 when you can find someone who wants to part with the rig.

FINALE

The commercial and kit gear we have just covered is not the entire spectrum of what is available. No doubt, experienced QRPers who read this section will wonder why I have not included a certain make or model of kit or commercial rig. The answer is simple—there is just not enough room. What I have tried to do is to cover the most readily available rigs on today's new and used market. The *QRP Quarterly*, local and regional clubs and the internet reflectors have all sorts of discussions about these rigs along with some of the other radio equipment available to the QRPer. **Appendix C** has the addresses and internet URLs of the various manufacturers featured in this chapter.

CHAPTER 5

QRP Operating Strategies

In this chapter we will explore some of the advanced operating techniques needed to be a successful QRP operator. The information is designed to sharpen your operating skills and start you on the way toward being a Master QRPer. In the process you'll qualify for DXCC, along with many other awards offered by the ARRL and other ham radio organizations.

The most prominent skill you can develop as a QRPer is the skill of listening. As my Grandfather George was fond of saying, "Dick, God gave you two ears and only one mouth so you can listen twice as much as you talk." Learning to listen for subtle telltale signs and nuances in a QSO or band conditions will speed you on your way to becoming a Master QRPer. Two other traits — tenacity and patience — go hand in hand. As we'll see later in this chapter, they can pay big dividends for the QRP DXer.

CW VERSUS PHONE

Just as CW has its special operating strategies, so does QRP phone operation. Once again, it boils down to a 13-dB disparity between your QRP transmitter and the rest of the wolf pack, running, at minimum, 100 W. There are certain things that you can do to greatly enhance the chances of getting into the other station's logbook. While some of these may be a rehash of topics covered earlier, it never hurts to reemphasize key elements of proven operating habits. The techniques used for CW operation are equally applicable to phone operation. Therefore, we won't spend the time separating these two modes, but rather cover the operating strategies for both modes simultaneously. The overall goal is to boost your operating skills far beyond the level of the average ham radio operator. That is, after all, the *real* secret to successful QRP operation.

BAND SCANNING

Let's go back to basics. Listen, listen and then listen some more. Trying to use the brute force method of barging through a pile up with only 5 W of RF is not the best course of action. If you listen to how the DX station is operating, the call areas they are working into, how propagation is affecting their signals and the signals of those they are working, whether they are taking all comers or going by call area number as well as whether or not they are working up and down the band between contacts, you'll greatly increase your chances of successfully working the station. This applies to CW as well as phone.

BAND SCANNING, THE HOME VERSION

Learning the idiosyncrasies of each band takes a lot of effort and time. Nothing takes the place of time at the radio. Listening to each band, being able to identify propagation characteristics — when the band opens and closes, how it reacts at various times of the year — makes for some fascinating detective work. The QRP DXer must know these things.

Band scanning is a method of milking the most out of each band by starting at the lower band edge and *slowly* working upward in frequency, stopping to listen carefully to the stations that are present. This is an on-going exercise in listening, which not only sharpens your listening skills but also teaches you a lot about the conditions of each band.

Band scanning lets you get the feel of each band, what the propagation characteristics are like and who is talking to whom. If you are in need of a specific country, and during your scanning exercise hear a station on the air from your target country, but that station is not working any stations in your geographic area, it is a pretty sure bet that you are not going to get the chance to work the needed station at that time. Note your target station's frequency and operating characteristics for future reference and comparison to published data in the various DX bulletins. This will make your chances of bagging him at a later date much better. If, on the other hand, this needed DX station is working other hams in your basic geographical area (let's say the east coast), then by all means, jump in and see if you can work it. You just never can tell what might happen!

DX OPERATING TECHNIQUES

The only way to work DX is to get on the air. This is something we've touched on earlier, but the simple fact remains that in order to work DX you must be in front of the radio. You won't work any DX by watching the packet cluster!

Previously, we talked about band scanning and developing your listening skills. When you listen, check out the signals for specific sounds. DX stations all have unique sound anomalies. Look for flutter, language accents, QSO style and content and, in CW, a particular "fist" style. A fluttery signal is most likely coming over the pole; therefore, it must be DX. It is relatively easy to spot a foreign accent and pinpoint a geographical area of the world. Sometimes the DX operator will have a unique accent but live in a different part of the world other than what would be expected from his accent. Don't be fooled, this type of DX op is still worth pursuing.

Listen, also, for conversational clues. If you hear a DX op talking about the weather on Heard Island, well...you *just might* want to hang around on frequency and try to work that station. Should you tune across a DX op running stations and giving out only signal reports, you might want to listen around the frequency to get an ID and QSL information and possibly pick up a new country in the process.

When stalking a particular DX station, knowing the sending habits of the operator (in CW) or his voice characteristics (on phone) is a very valuable piece of DX intelligence. Anyone who has worked Martti Laine, OH2BH, on one of his famous DXpeditions, knows that Martti's voice and operating style are distinctive. Therefore, it is easy to ID Martti when he is on the air. This could signal a new country being activated by a DXpedition. The CW "fist" of a DX op and associated operating characteristics can also speed identification of the station. All this preparation yields more productive time at the radio. Wasting time and not working your share of DX is counterproductive.

Remember the old saying: "Timing is everything?" With

DXing this is doubly true. Pick your operating times to coincide with the times that the target DX station is going to be on the air. It does no good to be scanning the bands hunting DX when the DX op is in bed at 0300 local time at his end! Also be aware of national holidays and major sporting events in your target country. Should the DX country's soccer team be playing in the World Cup, I doubt seriously that you are going to scare up any DX operators during the match!

As long as we're talking about timing, let's not forget the daylight-to-darkness and darkness-to-daylight transitions that provide some very good propagation paths into DX areas of interest. Called "gray-line propagation," this propagation enhancement can result in a drastic increase in signal strengths over the light-to-dark transition paths between two DX stations. QRP DXers should learn all they can about gray-line propagation since it is a great equalizer, allowing a low-powered QRP signal to be heard in areas that would be almost impossible using normal propagation paths.

One final thought on time; stagger your operating schedule to include weekdays and late night hours in order to snag some DX. By varying your operating times you expose yourself to more DX opportunities and you can avoid the huge wolfpacks that produce band-killing QRM.

Learning when and where to move your DXing operations can net some great DX catches. When the DX station ceases operation on one band, many DX ops immediately move to another band and commence operations again. Sometimes the DX op announces this fact. You need to be ready for an instant QSY to the new band and frequency so you can hopefully snag the DX before the wolfpack arrives. Many times the DX station does not announce a band change but merely stops operating. A rapid search on another band or two usually finds the station calling "CQ" with only a couple of stations calling him on frequency. Knowing the current propagation situation on adjacent bands can be a godsend in getting on frequency and bagging a new one when he QSYs.

"One-shot kills" when working DX are rare, especially for the QRP operator. Knowing this, take time and listen to the ebb and flow of the pileup. All pileups have a rhythm. Look for it. Watch how the DX station works the pileup. This is really important, as the ebb and flow of the pileup will offer you a chance

Listen, Listen, Listen, And A Dash of Luck

Randy Rand, AA2U, tells a funny story about trying to work a C21 (Nauru) station. Randy states that the C21 operator had a world class pile up in progress. Many of the stations in the pile up were so intent on transmitting that they were making a mess of the frequency. The C21 operator could not make out any call signs in the ongoing melee, so he stopped operating. Randy continued to *patiently listen* (there's those words again) on the C21's frequency and waited until the wolfpack had departed. He then gave the C21 a single call: "AA2U QRP." Nothing.

In about five minutes the C21 station started calling "CQ DX" again and once again the wolfpack descended, creating utter chaos. Randy continued to listen. Again, the C21 operator could not pull any one station out of the pile up so he looked down his list and called Randy! The morale of this story is read the pile up, listen, listen (and don't mindlessly transmit) and be patient!

Certainly luck played a part in AA2U's success, but so did Randy's highly developed operating skills. Had Randy been calling the C21 station mindlessly, like the rest of the wolfpack, he would not have heard his own call sign! One final point, many DX operators keep a secondary list of call signs as they hear them and when copy becomes tough on their end, they will refer to this list. Randy was in the right place at the right time and, through the use of his developed skills, bagged a very rare DX station while the rest of the wolfpack looked on. Good job, Randy!

to slide in your call sign with little or no interference (QRM) from other stations on frequency.

If the DX station is working split — listening on one frequency and transmitting on another frequency — it is vital to spot where he's working the stations that are calling him.

Often a DX op will say "UP 5 to 10," which means that he will be listening above his transmitting frequency by 5 to 10 kilohertz. Now, your job is to find out exactly where in that 5 to 10 kHz portion he's at! The best way is to toggle between his transmit frequency and the receive frequencies, using dual VFOs (if your rig has this feature). By flopping back and forth, you can track who the DX station is calling and, hopefully, find that station talking back to the DX station. Once you finally find where the DX station has just contacted the last station he worked, continue flipping between the DX station's transmit frequency and the frequency of his last contact to see if another station close by is his next choice. Many times the DX station will move a bit, either up or down, from the last contact to keep the calling stations spread out. Once you spot his modus operandi, track him for another couple of contacts and then place your transmit frequency where you think he'll be listening next. This all sounds very complicated, but once you do it a couple of times, it will become second nature when trying to work a DX station that's using a split-frequency technique.

Always, always, *ALWAYS*, give your full call sign *EVERY* time you call a station on phone and CW, too. This is especially true when trying to work foreign DX. A partial call will only cause the DX station to request your full call sign later in the QSO. Incidentally, this is one very good reason for a short vanity call sign! As long as we are on the topic of call signs, always use standard phonetics. While "Kilowatt Seven Sticky Zipper" is cute and gets an occasional laugh (especially from YL ops early on Sunday mornings during Field Day) it is not in keeping with the established phonetic alphabet. If the station you're attempting to work is a foreign DX station, and he has a limited command of the English language, well, you can see where this is going. Stick with the published standard phonetic alphabet and you'll be better off.

Try tail-ending the station that the DX operator is working, just as this station is signing off. Your 5-W signal won't seriously degrade the other station's signals, and he is ending the QSO anyway, so it's always worth a shot. Here, as with the ebb and flow of a pileup, timing is everything. Practice tail-ending but don't over due it. This is a proven operating technique and will definitely increase your chances of getting into the other station's logbook. Develop your tail-ending skills and you success with QRP with flourish.

QRP DXCC

The object of DXing is to work as many countries (approved DXCC entities as listed on the ARRL's DXCC list) as possible using only 5 W or less of radiated RF power. While there is no official QRP endorsement to the ARRL's DXCC award, they do offer a QRP version of the award for working 100 countries using QRP. See **Figure 5-1**. This is a "one time" award with no endorsements beyond the initial 100 countries. The QRP ARCI club, however, offers DXCC-QRP awards, with endorsements, based upon the ARRL's DXCC list. Of course you can earn the "full" ARRL DXCC award using only QRP. Your certificate (or plaque) may not say QRP, but you will know! The thrill of completing a DXCC award and adding countries to your total is hard to beat. You might even complete the Five Band DXCC challenge using

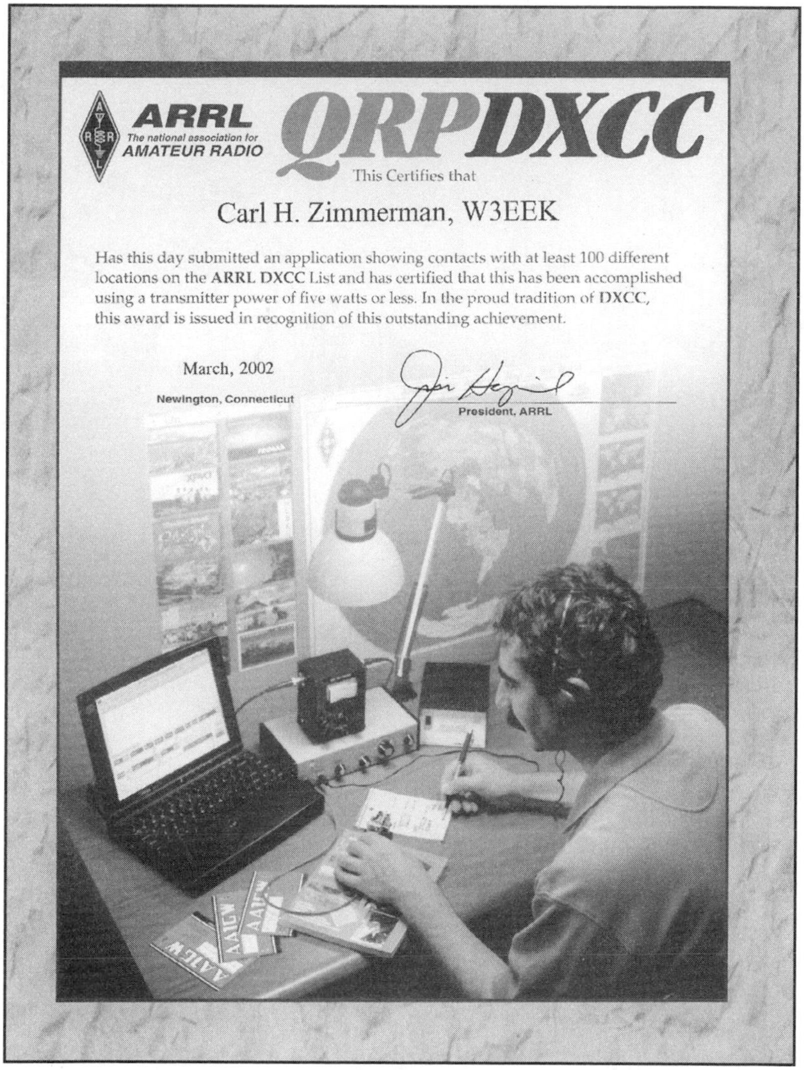

Figure 5-1 — To recognize the significant effort involved in completing and confirming contacts with 100 different DXCC entities, ARRL offers the QRP DXCC Award. To earn a QRP DXCC, you must use a transmitter power of 5 watts or less. The other stations do not have to be using QRP, however. Contacts with high-power stations count towards the Award, as long as you are using QRP. The QRP DXCC Award does not list the operating mode or band, and cannot be endorsed for additional contacts.

QRP. **Figure 5-2** shows a Five Band DXCC Certificate.

The first 25 to 50 countries are so easy to bag it is pitiful. When you pop over the 50-countries-worked mark, things will slow down a bit. As you continue to progress toward the coveted DXCC mark of 100 countries, you will need to refine your operating techniques, tactical strategies and station equipment to ensure success. Patience and tenacity play an important role in QRP

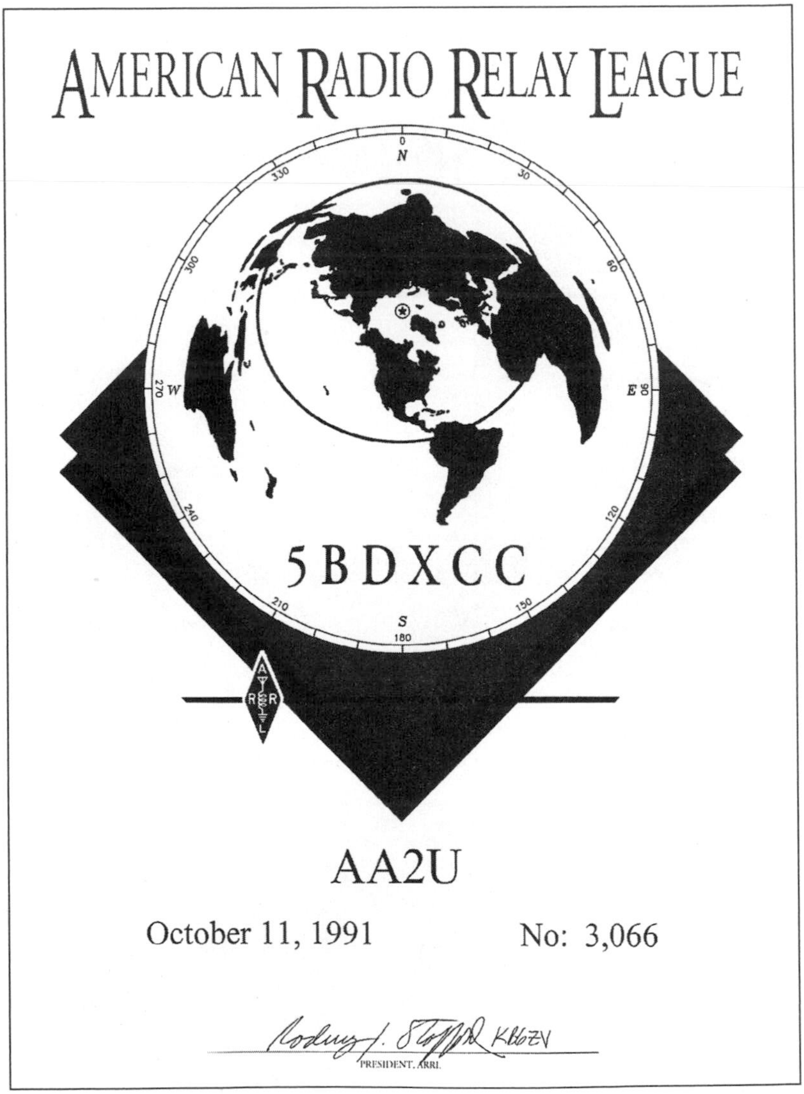

Figure 5-2 — The ARRL DX Century Club Award is the most presti-gious DX award available. Because the basic QRP DXCC Award cannot be endorsed for bands, modes or additional contacts, lots of QRPers earn the "full" award. Completing the 5 Band DXCC Award is no small task, even with high power. Earning it while using QRP for all contacts is an accomplishment of which you can be very proud!

DXing. Sooner rather than later you will realize that not every station will answer nor all pile ups be broken: QRP does have limits. Those times when you do snag a rare DX country, nail a new one, bust a big pileup or snag some DX tidbit from the jaws of the wolfpack make the effort all that more rewarding.

In keeping with our overall strategy, to be a successful QRP DXer does not mean you have to go out and spend $35,000 on a tower, rotators and several directional antennas. Quite the contrary, DXCC can be accomplished using nothing more than an old Ten-Tec Argonaut 509 transceiver and wire antennas. Of course, a rotatable directional antenna would go a long way to leveling the playing field with the Big Guns, but it certainly is not a prerequisite.

SETTING GOALS

One way to chart your progress and to hold your interest in QRP DXing is to set some *realistic* goals. Probably the easiest goal to think of is the award itself. Your first goal might be mixed-mode DXCC. By mixed mode we are talking about both CW and phone contacts. With the current crop of computer logging/awards tracking software available to the QRPer, it is a simple matter to track your progress on CW, Phone and mixed-mode DXCC simultaneously. Reviewing your DXCC standing periodically will help focus your efforts and increase your determination to qualify for the award.

No goal is complete without a time frame, so let's assign an arbitrary time to accomplish mixed-mode DXCC. Perhaps you want to reach that level in 6 months. This is definitely a "doable" goal, given several hours of operating time each week along with a couple of world-wide DX contests blended into the mix. Contests, by the way, are a great source of DX fodder and can go a long way toward boosting your DXCC countries-worked list. Don't be afraid to get into the fray and work some new ones. If you are unable to dedicate that much operating time to the goal, then modify the time frame to say, 12 months. Even on a highly erratic operating schedule, you should be able to work the required 100 countries in a year. Getting the necessary QSL cards may take a while longer, however!

These same goal-setting strategies can be applied to QRP contesting, too. When you enter a world-class contest that offers a QRP category, you will only be competing against other QRPers. Few of us can hope to successfully win against some of the top contest operators but we can set goals for each contest we enter. For instance, if you operated 24 hours of a contest the previous year, try increasing your operating time. Failing that, try to work more stations ("Increase the rate!") in the same amount of time or less. Track your scores for each contest and work to improve them on subsequent contests.

As you approach your initial DXCC goal, start thinking about "lifting the bar" and extending your basic goal. How about 200 countries in under 12 months? Along with this goal modification comes the realization that you should start seriously thinking about adding a directional antenna of some form to your antenna farm.

Big towers and big antennas take big money! If you don't have the budget or lack the space for a large tower and directional antenna, then think small. How about a small tri-band Yagi beam on a 30 foot push-up mast attached to the side of your house? Hamfests are a great source of used Yagis and vertical antennas, so shop around and put something small up in the air to give your QRP signals some directivity.

Verticals are especially suited to DXing when properly installed with a good radial system. While a quarter-wave vertical antenna exhibits omni-directional radiation and reception, it tends to have a much lower take off angle for the RF wave, which means that the RF energy travels farther before encountering ionospheric bending. Verticals are a lot easier to conceal in populated areas where zoning restrictions might apply.

If a tower and rotatable beam are out of the question due to finances, physical space limitations or zoning/rental covenants then modify your extended goals to include achieving single band DXCC on one or two bands of your choice. There are many ways to chase DX and it doesn't take a lot of time, money or antenna hardware to do it.

THE CHASE

The act of searching, stalking and bagging a new country is the essence of DXing. Nothing beats "The Chase!" Developing a strategy to make the chase more successful is one of your primary duties as a QRP DXer.

Start this process by assessing the rarity of each DX country. Some DX entities have thousands or tens of thousands of license hams who are active on the air. These are the *common* DXCC countries like Japan, Canada, Germany, Australia, the United Kingdom, France, most of Europe, and so on. Other DX entities have far fewer active hams and we can classify these countries as *uncommon*: Central and some South American countries, along with many of the African countries. The remaining DX entities, having very few or no active ham population, are classified as *rare*.

A little detective work in the *DX Callbook* will yield a list of common and uncommon DX countries. To separate the uncommon from the rare DX entities, you will have to consult the Top 100 list as compiled by *DX Magazine*. This Top 100 list is compiled from the most-wanted lists of many active DXers throughout the world. The lists rank the standing from number 1 (most rare) to number 100 (merely hard to find). Once you classify your quarry by its rarity, you can start developing a strategy to get that country into your logbook.

When starting out in the DXing game, it makes more sense to go after the common countries since there are more active hams in those countries and their signals should be much better than some of the uncommon and rare countries. So, our first goal is to bag 100 countries from the common countries list and qualify for the basic award. Once that has been done it will be time to concentrate on the uncommon and rare DXCC entities. As you start working some of these less populated countries, you'll notice that their signal strengths are down, QRM from the wolfpack is more intense (this increases with the country's rarity), and the chances are the DX operator's skills are not on par with what you are used to working with. These stations will be more difficult to work and it would not be uncommon to spend several weeks getting a rare country into your logbook.

During this phase of your DXCC quest you should try to balance the ratio between common, uncommon and rare DXCC entities. As your country count nears the 300 mark, you will be left with only the more rare countries. In a nutshell, you will have worked all but the areas of the world that have no active ham population unless there is a DXpedition or guest operator who activates that location for a brief period. At this juncture, it is not uncommon to measure the times between working a new country in months! The main problem is now to remain active, always looking for that sudden activation of a rare one that you don't have. Scan the DX packet cluster and the Internet DX newsgroups for information on upcoming DXpeditions or possibly new operators in these rare areas.

THE DX CLUSTER

The DX packet cluster can be a useful tool for the QRPer, but not the way you'd expect. Clusters are used as spotting tools for DXers. Whoever hears or works some DX posts the frequency, DX station's call sign and any other pertinent data on the cluster, where this information is read by anywhere from a few to several thousand stations who are also watching the cluster. A rare DX station shows up, gets posted on the cluster, and watch out: instant pileup! Obviously, this portion of the band is where you want to steer clear of. Charting the operating characteristics of DXpeditions over several days via the cluster can result in your being on frequency when the DX operation fires up on a new band. Getting a clear shot at the rare DX station in this manner greatly enhances your chances of getting into his log.

Although I occasionally use the DX cluster, I use it to stay away from the pileups. Actually, by watching the information scroll up the screen, the adroit QRP operator can get a picture of how the band is shaping up and an estimation of the chances to work various areas of the world. Most clusters also show the last several WWV/WWVH propagation broadcast data, so you can keep abreast of the latest propagation data and further maximize your operating time by knowing which bands should be open.

FOOD FOR THOUGHT ON DXPEDITIONS

It's a tough job but somebody's gotta do it! Thankfully we have a contingent of world class DX operators that regularly band together and activate a rare country or DXCC entity so the rest of us can "snag a new one." While DXpeditions allow you to bag otherwise unattainable DX entities, the task of putting together the funding, personnel, equipment and transportation to a rare DX spot is formidable. Many of these DXpeditions are years in the planning before a single QSO is made. Martti (OH2BH) Laine's book, "*Where Do We Go Next?*" is excellent reading and provides some great insight into exactly what takes place before, during and after a world class DXpedition. These "little" junkets are expensive; *real expensive*. The operators chosen for these DXpeditions are some of the best DX ops in the world. Their equipment and antennas are world class, too. This is good news for the aspiring QRP DXer. These operator's skills are unbelievable! They can pull call signs out of a howling pileup or down the noise that most people cannot even hear! Your chances of working a rare country by working a DXpedition is really quite good. One hint: Wait for the initial pileups to die down. This usually happens after the second or third day of the DXpedition.

Many DXpeditions also maintain an Internet website where the latest operating schedules are posted along with other pertinent information regarding the event. Additionally, these website may have the latest uploaded loggings by the DX ops, and you can see if you are in the DXpedition's logbook around 24 hours after you make your QSO.

All DXpeditions can use your support. Many times the website has a method to provide for contributions to the DXpedition before, during and after the event. If you worked the guys on DXpedition, don't feel bashful about sending them a couple of bucks to help offset these tremendous costs of putting together a huge effort.

QRP CONTESTING

Many ham radio operators lament the dreaded "contest weekend" that turns otherwise calm HF bands into a tumultuous sea of churning RF energy. Actually, the big contests only occur a few weekends per year. Smaller regional, state QSO parties, and specialty contests are continuously underway.

For those who don't understand contesting or the psychological composition of a contester; the idea of spending 48 hours glued to your radio, swapping seemingly meaningless signal reports with other stations may seem a bit on the weird side. Looking at contesting from another perspective, a contest weekend is 48 hours of "high speed, no drag" ham radio fun! For those really into this aspect of the hobby, it is an extremely intense experience that brings out the competitive spirit of all who compete.

Using only 5 watts or less, how can a QRPer hope to compete against the world class contest stations with their kilowatt-plus signals and stacked antenna arrays? The answer: you don't. Instead you compete against other QRP stations who enter the QRP class. All of the major worldwide DX contests have a QRP entry class. You're still going to have to slug it out on the bands but your score is calculated only against other QRP entrants. Do not confuse the "Low Power" category with the "QRP" category. Low power, in this case, generally means up to 150 W RF output not 5 W!

Contesting is a great way to boost your DXCC totals. There are special contests DXpeditions that activate uncommon countries just for major contests. This is a boon to the QRP DXer, since it presents a chance to work some countries that otherwise might never get in the logbook. Many of the DX ops you will work are some of the best in the world. This greatly enhances your chances of getting a contact. Their listening skills are so well honed that they can hear grass grow!

The arrival of fall each year signals the start of the contest season. Therefore, use the summertime to improve or add to your existing antenna farm. Antennas are the key to big contest scores, whether you are using 5 W or 1000 W. Even if you don't have a directional antenna at 120 feet, you can still compete using wire antennas.

Goal setting is a must for the QRP contester. Realistic goals are the key. When I enter a contest my main goal is to better my last score. This, for me, is a realistic goal. While I'd love to work DXCC in a single weekend (and I know that it can be done using QRP power levels) I realize that my chances of doing this are pretty slim due to my limited operating time. Therefore, I set a realistic goal of 50, 60 or 76 countries, based upon my station limitations and operating schedule.

You cannot expect to turn in a great contest score without spending a lot of time in front of the radio (gosh, where have we heard that before?). The more time you spend on the radio the more contacts you make and the more your score improves. Many of us do not have the luxury of being able to spend an entire weekend at the radio and must choose our operating time carefully, dividing it between family obligations and work schedules. This is okay, too. You compete at what ever level you desire. After all, this is *only* a hobby, right?

What follows are some select tips on how to maximize your fun and contest score. These tips are the result of conferring with fellow QRP DXers/contesters and my own experiences over the years. Enjoy.

Read and understand the contest rules. As dumb as this sounds, it's nice to be on the air the same weekend the contest is running! Also, know the type of contest and the proper exchange. This definitely saves embarrassment.

Don't try to compete with the "Big Guns" and sit on one frequency calling "CQ Contest." It just won't work at QRP power levels. Instead use the "Search and Pounce" (S&P) method. S&P gives you the mobility to move about the bands and quickly grab QSOs and multipliers.

Work the loud stations first. If you don't succeed getting through after the first couple of calls, note the frequency (or enter it into the memory slot of your rig) and come back later and try again. Keep moving, constantly searching for new contacts. S&P also helps fight the boredom that sets in after a few hours.

Once you have worked through the loudest stations (called the "first tier" stations) go back and work the second tier and then the third tier stations. In this way, you thoroughly work each band and maximize the number of contacts.

Stay away from "Kilowatt Alley," at the bottom end of each band. You cannot possibly compete with these guys so don't even try. Instead cruise the middle and high portions of each band, netting QSOs as you go along (S&P). On the second day, toward the end of the contest, cruise through Kilowatt Alley and snag the monster multi-multi — multiple transmitters and multiple operators — kilowatt stations that are, by that time, begging for contacts.

Use a computer to log and keep score. There are many outstanding logging programs available to the QRPer. Most support the major contests. Find a good one that you are comfortable with and use it. Thoroughly learn the ins and outs of the program and use the same one every contest. Computer logging simplifies the record and score keeping (not to mention dupe checking) during a contest and at the end of the fray, you can pull your logs off in Cabrillo format to send in for scoring. Most major contests now require contestants to submit their logs in Cabrillo format to insure accuracy and reduce the time and efforts required to compile the contest scores.

Be aggressive. Attitude is everything. Act like you are running a kilowatt. That's right; don't be afraid to jump into the melee and slug it out. If, after three tries the other station is not in your log, move on and return later (S&P).

Regularly check other bands, especially the higher bands, as the day progresses. Many times, you will be rewarded with a band opening that will garner new Qs. If you hit a band just as it is opening, you won't be in competition with nearly as many "Big Gun" contesters, so your chances of success are even better.

Sharpen your tailending and listening skills. Listen for how the other operator is handling a pileup and then slide your call in jus as someone else finishes. This is a very effective technique and will net many Qs.

Use a memory or contest keyer for CW and a digital voice keyer (DVK) for phone contests. This takes a lot of work out of the exchange and it allows you to concentrate on identifying and working new stations without having to worry about manually transmitting the exchange.

Scan the bands 15 to 20 minutes *before* the start of the contest. Listen for other stations signing their call signs and making what seem to be mock contest exchanges. These operators are "stretching' much like an athlete does prior to exercising. Tuning the bands in this manner will give you a feel for propagation and get you off to a good start.

Start the contest on the highest band that is open into your area. If you don't establish a good QSO rate, drop down to the next band and give it a try. The object is to find the band that is propagating the best and gives you the best shot at making lots of Qs.

Keep your exchange short. Send the exchange only once. Repeat the exchange *only* if the other stations asks for a "fill."

Don't send only the last two letters of your call sign when calling another station. This is a waste of time, since the other op will need your entire call for his log. Use your full call sign each and every time you call another station.

In the event you are called by several stations, pick up the weakest one of the group first. This may be your only chance to bag this station should propagation suddenly shift or take an unfavorable dive.

Transmit at the fastest speed at which you can comfortably send. Since most contest exchanges are quite predictable, listen carefully to the high speed stations, get their info before you call them, then goose your keyer's speed up and complete the contact. This gets you into their log (and vice versa) and gives them the idea that you just might know what you're doing!

Get plenty of sleep prior to the contest. You cannot "make up" sleep. If you get tired during the contest, and you will, try to plan your sleep breaks so they fall at a time when the QSO rate will typically be low.

While rules vary with each contest, as a QRP entry you cannot generally use DX packet cluster spots during the contest. This places you in an "assisted" category and you will now be competing with higher powered stations.

Keep abreast of the latest contest developments by reading *NCJ: The National Contest Journal* and QRP newsletters, websites and internet reflectors.

Approach QRP contesting as a challenge and as a learning experience. This is a time when you can try new techniques and sharpen your DXing skills. Jump in and have some fun.

AND FOR THE PHONE OP...

QRPers like making CW contacts. Why, because CW is the preferred mode when working QRP. In actuality, CW has over a two to one advantage over SSB due to the narrower bandwidth and the ability of the human ear to pick out the CW note more readily than voice frequencies out of the noise. This is not to say that phone or data operation has no place in the QRP arena. Quite the contrary, many times QRPers are driven to try SSB simply for the challenge.

QRP and phone operation are not mutually exclusive terms. On the contrary, many QRPers shy away from phone operations, mistakenly thinking that using phone will lead to much frustration and wasted time. While there are some distinct advantages to the use of CW (the mode of choice for QRP communications) phone operations should not be ignored or side stepped. Sometimes, the only way to log a rare DX station is by using phone.

This chapter will outline some specific things that the QRPer can do to enhance the chances of being successful using phone as well as focusing on advanced operating skills that the QRPer needs to develop.

Of particular importance, when we discuss phone operation, is how we can control our transmitted audio so our QRP signals will be heard by the distant station. I will relate my personal quest to add a studio type microphone to my station along with some audio engineering procedures that anyone can duplicate to get that "perfect" audio sound out of their transmitter. Finally, we'll cover some operating tips just for the QRP phone operator.

CW RULES? NOT HARDLY!

All too often QRPers (myself included) abandon phone operation in favor of CW. Without a doubt, CW is more efficient than single sideband[1], but that's not to say that QRP SSB operation is not possible or practical. The problem with successful QRP phone operation is one of perception and acceptance. Many times QRPers speak in irreverent terms regarding SSB. It seems, to some, that QRP and phone operations are mutually exclusive terms. This is definitely not the case.

The gravitation toward CW is directly related to our comfort level. QRPers like CW because we are confident that we will be successful using it. The narrow bandwidth of the CW signal coupled with competent operator skills equates to success. QRP phone operation is an altogether different ballgame. Voice intelligence of the phone signal can become lost in the band noise.

While CW ops will spend lots of money and time selecting and adjusting a set of paddles, very few hams will spend the same amount of time and money selecting a microphone for their HF station. After all, a mic is a mic is a mic, right? Sadly, audio engineering within the ham radio hobby has also largely been neglected by the average operator. When we leave the high power world and play in the QRP arena, we give up at least 13 dBW of power advantage. By examining our stations and applying some simple audio engineering, we can regain some of this power disparity. Couple this with some killer operating skills, and you'll be amazed at the DX you can work using QRP phone.

SOME AUDIO BASICS

Human speech is a complex waveform covering several thousand Hertz in the audio frequency portion of the spectrum. Communications audio is universally thought of as that portion of the audio spectrum between 300 and 3000 Hz. Therefore, we

[1]George Jacobs, W3ASK and Theodore Cohen, N4XX, *The New Shortwave Propagation Handbook*, pp 1 to 19.

are dealing with a fairly narrow band of audio frequencies when we want to produce good audio for the transmitter circuitry. Looking a bit more closely, speech *power* is centered between 200 and 600 Hz for the male voice, while the greatest speech *intelligibility* occurs around 2 kHz. Approximately 33% of speech intelligence is within an octave centered on 2 kHz (between 1.414 to 2.828 kHz) in a normal speech pattern. These are called "mid-range frequencies."

What's an octave? An octave is a term that defines a specific frequency change within the audio spectrum. The range of 500 Hz to 1 kHz is one octave. From 1 kHz to 2 kHz is another octave. In other words, an octave increase doubles the frequency while an octave decrease halves the frequency. Octaves are handy measurements when we talk about filters and audio equalizers (EQ units).

In the audio world, we deal with voltage gains/losses expressed in dB. The dB formula used with voltage is 20 log (E_1/E_2). In audio, where everything is referenced to a voltage level, every *6 dB* yields a doubling or halving of the signal. Therefore, if we have a perfect audio filter that has a 6-dB roll off per octave at 1 kH, at 500 Hz we should see one half the input signal voltage and at 250 Hz only one quarter of the input signal.

We live in an imperfect world. This applies to filters, too. A real filter's cutoff frequency (F_{co}) is measured at 3 dB down from the center frequency (F_o) of the filter. This is *not* a linear roll off. The linear slope of the roll off occurs even further away from the center frequency, resulting in a much wider passband. This is the reason that we need to use several poles of filtering to achieve the desired passband characteristics in our audio filters and EQ units.

Audio equalization is the key to successful QRP phone operation. By using properly designed high and low pass filters we can tailor the microphone toward the desired communications audio response curve in pursuit of that elusive "perfect" audio. This equalized audio, when applied to the transmitter, will yield a much more intelligible SSB signal, with a lot more audio "punch." This is because the signal voltage in the maximum intelligence portion of the communications audio range makes up a larger proportion of the transmitted signal. While we *are* "processing" the audio we *are not* doing anything to alter the RF power output of the transmitter like some audio processors claim to do. Audio equalization can give us the competitive edge in

QRP by improving the *intelligibility* of our low power signals.

How we go about manipulating the microphone audio is, in some cases, amazingly simple. An expedition to my local Radio Shack yielded a general purpose dynamic, omni-directional mic (RS # 33-3030) on sale for under twenty dollars! I bought it and headed back to the shack to do some serious testing. The documentation revealed that the mic had a fairly flat frequency response form 100 to 12,000 Hz. This is fine for use on a PA system but not what I needed for my phone station.

An article by Bob Heil, K9EID, described a mic equalizer and also offered some tips on how to enhance microphone characteristics.[2] I decided to try them on my new mic just to see what would happen. An on-air test with Fran Slavinski, KA3WTF, proved most enlightening. He listened to the stock mic and commented that the low frequency end was very pronounced yielding a "bassy" quality to the audio. I tried Bob's trick of wrapping the sides of the mic head with a layer of tape and had Fran listen again. This time the results were much improved. Fran said that the low frequencies were greatly attenuated and the mid-range frequencies were coming through nicely.

I took this modification one step further by cutting a three inch piece of light card stock, 1.5 inches in width and placing it around the *inside* of the mic head. Removing the tape from the outside of the head greatly improved the cosmetic appearance of the mic and still sealed off the side slots, effectively closing off the air chamber inside the mic head. With the air flow behind the mic element restricted, the low frequency response is attenuated, which allows the mid-range frequencies to become more dominant. Instant equalization! Like I said earlier, some things are *real* simple.

One other "simple fix" is to place a 0.01-µF disc capacitor in series with the hot mic lead. This capacitor, in combination with the impedance of the mic element, creates a high pass filter that further rolls off the low end response at or below 300 Hz. Using these two modifications transformed the RadioShack mic into a very useful addition to my shack with minimal outlay of cash and effort.

[2]Bob Heil, K9EID, "Equalize Your Microphone and be Heard," *QST* July 1982, pp 11 to 13.

A COMMERCIAL EQ UNIT, THE MX-604A

Since I chase my share of DX, I wanted to further alter the microphone characteristics to enhance those frequencies from 1.5 to 2.5 kHz. At the urging of Bob Heil, K9EID, of Heil Sound (**www.heilsound.com**) and George Baker, W5YR, who both shared their thoughts on audio equalization with me, I purchased a Behringer MX602A mic mixer/equalizer. A trip to the Behringer website (**www.behringer.com**), will yield the specs on this unit. Street price is around $50, down form about $125 several years ago. This unit is a great buy for the frugal QRPer desiring to improve his audio for phone operation.

It should be noted that Bob Heil's philosophy regarding audio engineering for ham band phone operation is to "Keep It Simple, Stupid!" He revealed that since we were only talking about a couple thousand Hertz of audio passband, you don't need a huge multi-band or even a 5 band EQ unit. All you really need is a 3 band equalizer to alter the mic audio for near-perfect SSB audio. Enter the Behringer model MX602A. The Behringer unit can take up to four mic inputs and has some basic equalization built into the unit, so you can alter the mic audio at will. At a around $50, the MX602A is money well spent. Using the MX-602A, I was able to adjust the transmit audio on both the high and low end to achieve a good peak in mid-range performance around 2 kHz along with a substantial roll off at 300 and 3000 Hz.

Accepting the challenge to work more QRP phone contacts, I started out with my new mic and promptly worked the following DX: VK3, TG9, ES4, LA1, JT1, IK3, ON7, LZ2, EA6, OD5, OM5, 4X4 and 9Y4. All these stations were worked at 5 watts or less on a very erratic operating schedule over a two week period. QRP phone operation is a BLAST!

CHAPTER 6

Antennas for QRP

Antennas are the life-blood of the QRP operator. This is the one area that can easily be manipulated to vastly improve the QRPer's on-air signal. In this chapter we'll give you an overview of the more common antennas used by QRPers. We will also look at several antennas that are simple to build or easy to procure and which will enhance your station. Since portable/mobile QRP operating has become so prevalent, I've included a special section devoted to portable antennas for use in the bush.

Earlier we touched on the three things that make up a QRP station: a radio, an antenna and your operating skills. We have covered the first and third in detail, so now let's concentrate on antennas.

While all three of the major components of a QRP station are important, the antenna(s) used by the QRPer are what makes the difference on the air. Your skills can be world class and your rig can be the latest marvel of electronic technology, but unless you have a good "skyhook" you won't talk to many people.

Wire antennas are, by far, the most popular type of antennas in use by QRPers. They are inexpensive, relatively easy to erect, perform very well when properly erected, and are low profile when compared to rotatable arrays. In this chapter we are going to look at the most common wire antennas and give you an idea of how to implement them at your location.

END-FED WIRE ANTENNAS

The end-fed wire is the simplest of all antenna designs. It consists of a length of wire hooked to the output of the rig or antenna

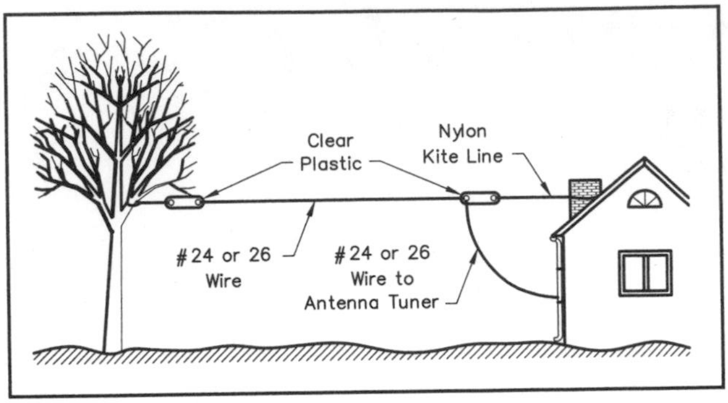

Figure 6-1 — An "invisible" end-fed antenna in your backyard can go virtually unnoticed.

tuner, which goes out the window up into a nearby tree or possibly hung down the side of a multistory building. End-fed wires are easy to erect and offer mediocre performance. **Figure 6-1** shows a typical home-station installation. They are field expedient and often the choice of backpackers and hikers.

One end-fed wire can offer multiband operation provided the wire is at least $^1/_4$-λ long at the *lowest* operating frequency. This would be about 67 feet for 80 m or about 33 feet for 40 m. It would be difficult to match a wire cut for 40 m operation on 80 m. It's just too short. You can use an 80 m end-fed wire on 40 m and all the shorter-wavelength HF bands, however, with reasonable results.

Possibly a better way would be to pick some non-resonant length that is not harmonically related to any of the bands and still over $^1/_4$ λ (67 feet) on 80 m. I have used a 90-foot length of wire as the main radiating element of an end-fed design with great success for several years. Ninety feet is approximately 0.64 λ on 40 m, which is about right for one half of an extended double Zepp design I use at home (more on the Zepp design later). It is nonresonant on the ham bands and long enough on 80 m to fill the requirement of being longer than $^1/_4$ λ on that band. It loads up just fine on all the HF bands.

One critical thing about using an end-fed wire is to be sure you also include a counterpoise wire hooked to the ground side of your tuner or back of the rig. The counterpoise wire(s) must be $^1/_4$-λ long on *each* band. This means you must have a counterpoise wire for each band, hooked to the back of the tuner. This counterpoise makes up the other half of the antenna and serves to capture antenna currents and improve radiation resistance of the wire antenna. A very good multiband counterpoise can be made using

Figure 6-2 — This diagram shows how you can cut a length of multiconductor rotator control cable to make a multiband counterpoise for an end-fed wire antenna.

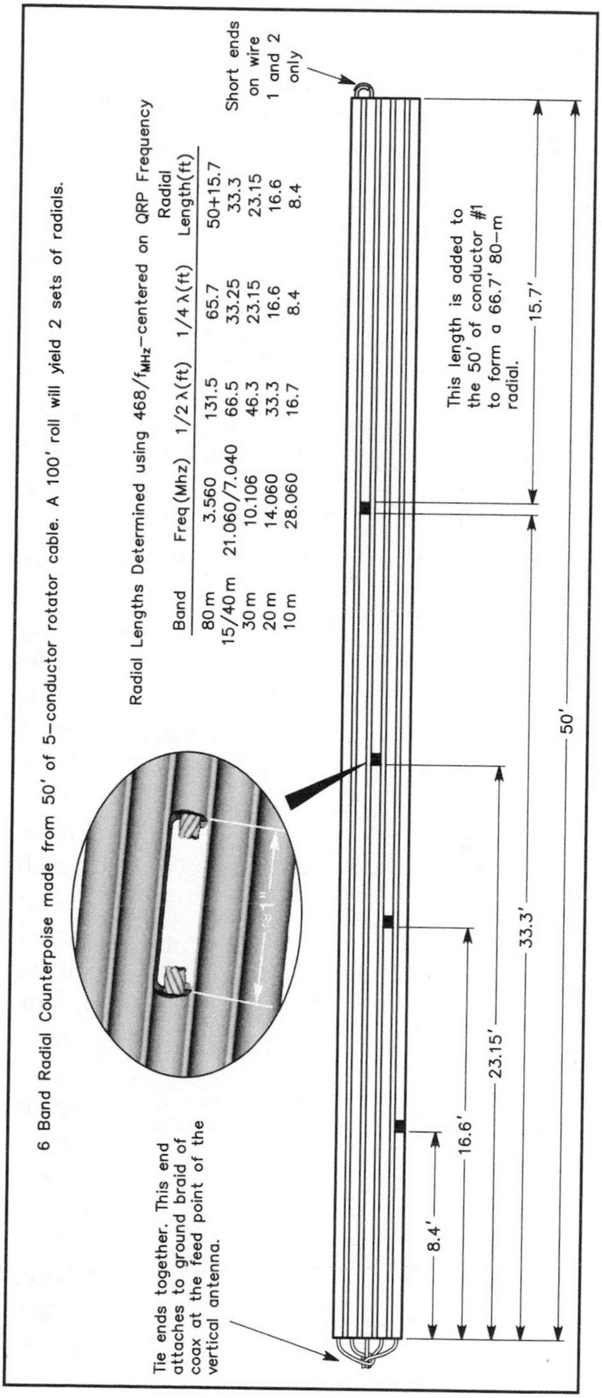

Band	Freq(Mhz)	1/2 λ(ft)	1/4 λ(ft)	Radial Length(ft)
80 m	3.560	131.5	65.7	50+15.7
15/40 m	21.060/7.040	66.5	33.25	33.3
30 m	10.106	46.3	23.15	23.15
20 m	14.060	33.3	16.6	16.6
10 m	28.060	16.7	8.4	8.4

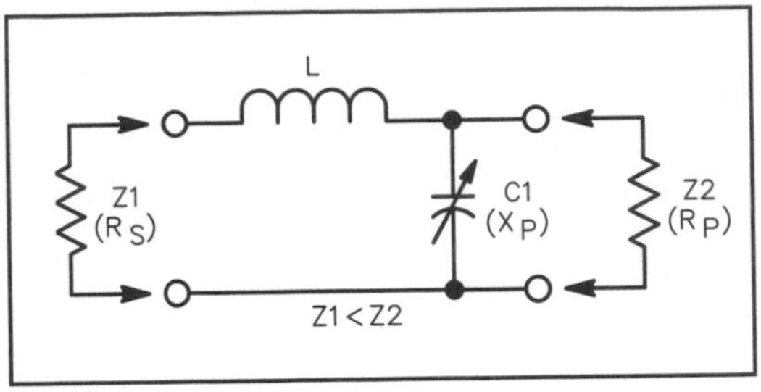

Figure 6-3 — This drawing shows an L matching network for use with an antenna system. The lower of the two impedances to be matched, Z1, must always be connected to the series-arm side of the network. The higher impedance, Z2, must be connected to the shunt-arm side of the network. The positions of the inductor and capacitor may be interchanged in the network for some matching conditions.

multiconductor rotator cable available at RadioShack. Check out the drawing in **Figure 6-2** for details.

While I have used an end-fed $^1/_4$-λ resonant wire connected directly to the back of the transmitter, it is not an ideal situation. Some form of impedance matching is almost mandatory. Therefore, a simple antenna tuner is highly recommended. **Figure 6-3** shows a simple L-Match tuner that can be used to interface a 50 Ω transmitter into a random length of wire. Should the impedance excursions become excessive, you can always reverse the input and output ports and match a greater impedance range. L-Match tuners are very popular and extremely easy to build. More sophisticated automatic tuners are also available, such as the LDG Electronics unit shown in **Figure 6-4**.

Figure 6-4 — The LDG Electronics Z-100 Autotuner is a very handy device, especially for the ham on the go. This tiny tuner can match coaxial-cable-fed antennas. With the optional external balun it can also handle random wire antennas and antennas fed with ladder line. The tuner operates from a 7 to 18 V dc source and uses latching relays to save battery power. With 200 tuning memories, this unit will remember the proper settings for many tuning conditions. **(Photo courtesy of LDG Electronics)**

One variation on a theme is to suspend an end-fed wire away from the edge of your multistory building and add a weight to the end of the wire. Drop the wire out a handy window allowing it to drape down the side of the building. A broom handle makes a good insulated support to keep the wire away from the side of the building. Hook this wire into the tuner, add your counterpoise wire to the ground side and tune for maximum output. This design is especially attractive to the "antenna challenged" among us who must suffer with antenna restrictions or Draconian covenants.

DIPOLE ANTENNAS

In my humble opinion, nothing beats a dipole antenna for simplicity, ease of installation and excellent all around performance. *The ARRL Antenna Book* has a ton of information regarding building and erecting dipole antennas. See **Figure 6-5** for

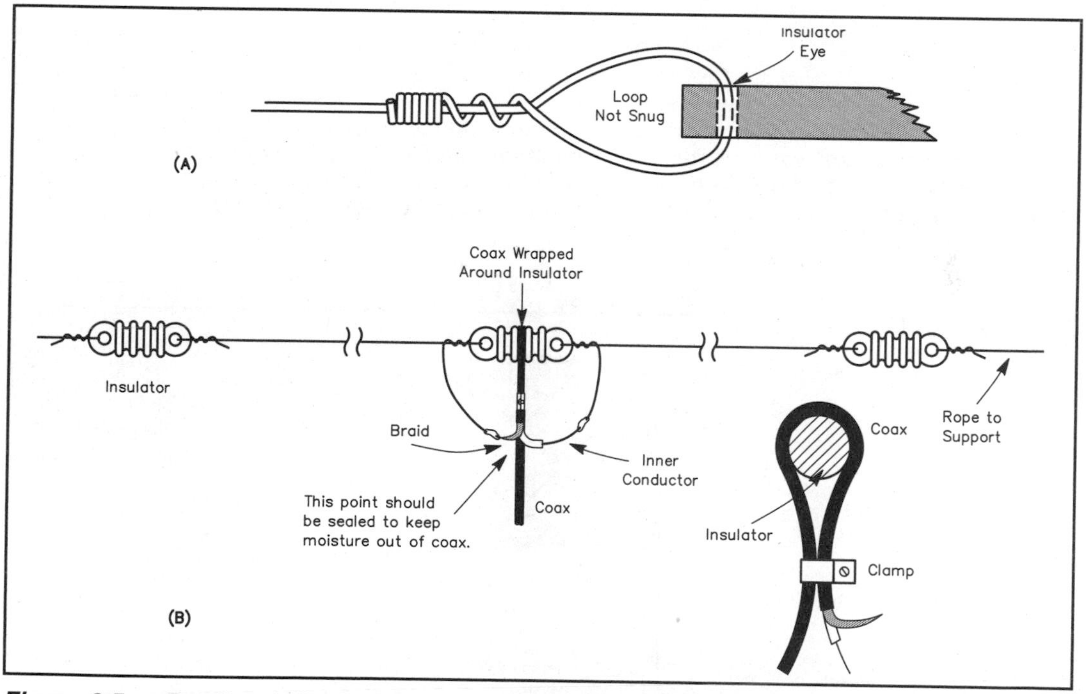

Figure 6-5 — Details of dipole antenna construction are shown here. Part A shows the end insulator installation. B shows the completed antenna. While there is no balun shown at the feed point, one is often used with coaxial feel line because the dipole is a balanced antenna, but coaxial cable is an unbalanced feed line.

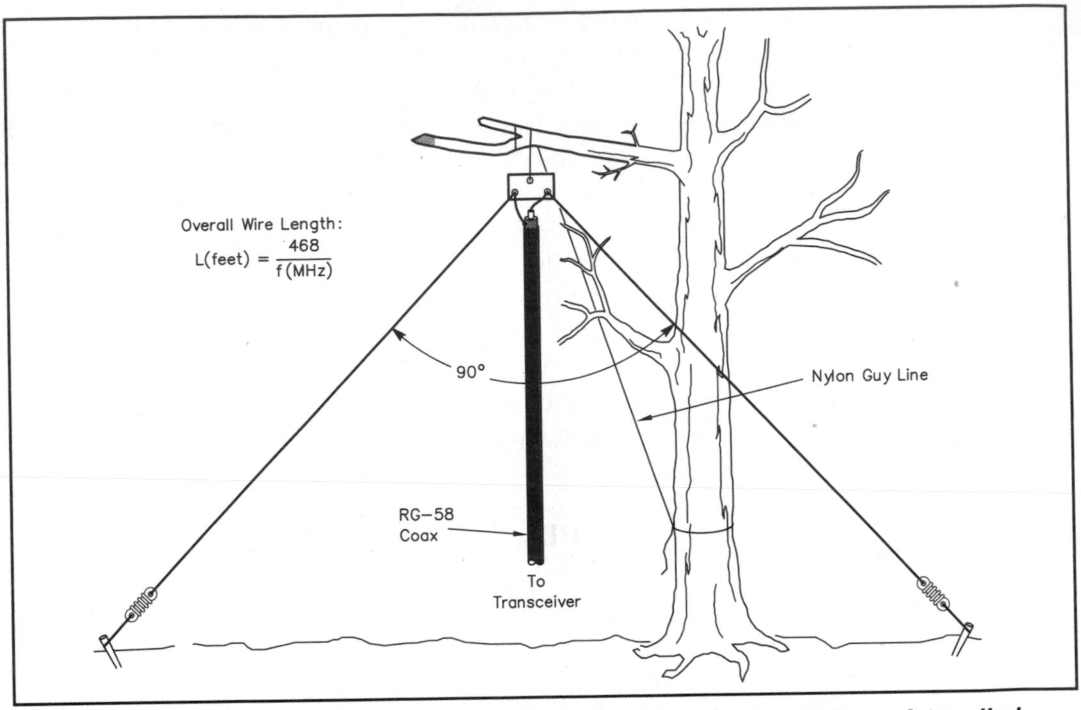

Figure 6-6 — Here, an inverted-V dipole is configured for single-band use. A tree limb provides a convenient support for the antenna center. RG-174 miniature coaxial cable may be used in place of RG-58 if the length is kept under 50 feet. (RG-174 cable is very lossy, and is not recommended for 14 MHz and higher frequency bands.)

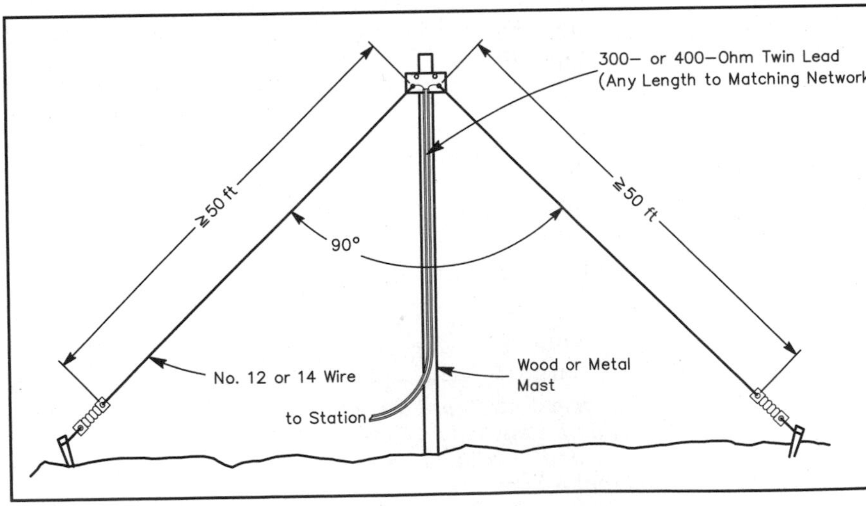

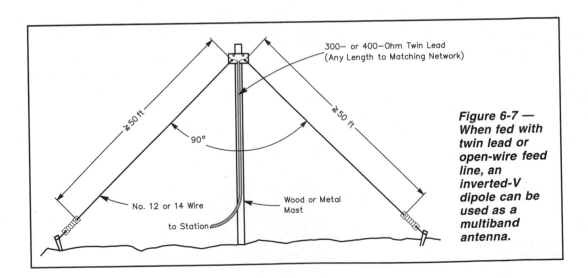

Figure 6-7 — When fed with twin lead or open-wire feed line, an inverted-V dipole can be used as a multiband antenna.

typical construction details. If you can afford the real estate, a full size dipole on each HF band is a very good antenna farm. Once you go to wavelengths shorter than 20 m, rotatable arrays become attractive, but you can still work a lot of DX with a dipole. Just remember to erect them as high and in the clear as possible. Additionally, dipole antennas, installed close to the ground, offer excellent high angle radiation, which means that they are great for local nets and ragchewing. They are also simple to erect as inverted-V dipoles, as **Figure 6-6** shows. This makes an excellent backpacking antenna. A combination of dipole antennas at various heights makes for a very versatile antenna farm. A single inverted V dipole fed with twin lead or open-wire feed line makes a good multiband antenna. See **Figure 6-7**.

THE EXTENDED DOUBLE ZEPP

One variation of the common dipole that I have become very fond of over the last year or two is the 40 m extended double Zepp. This is a dipole variant that is 0.64 1 per leg and fed with 450 Ω ladder line through an antenna tuner that employs a 4:1 balun. See **Figure 6-8**. The antenna is longer than a standard 80 m dipole but works very well on 80 m. It will load on 160 m and offers mediocre performance on that band. It really shines on 40, 30 and 20 m, providing gain and directivity. Check out the antenna plots in **Figure 6-9**. While this particular dipole antenna is large, the increased 30 and 40 m performance warrants serious consideration. We are talking wire elements, so the cost of material is low. The EDZ is a great antenna that is often overlooked by today's QRPers.

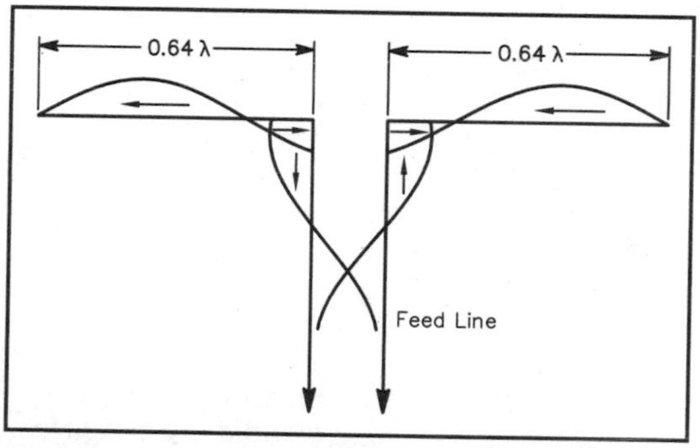

Figure 6-8 — The extended double Zepp antenna gives about 3 dB more gain than a ½-λ dipole antenna. This drawing shows the current distribution along the antenna and feed line.

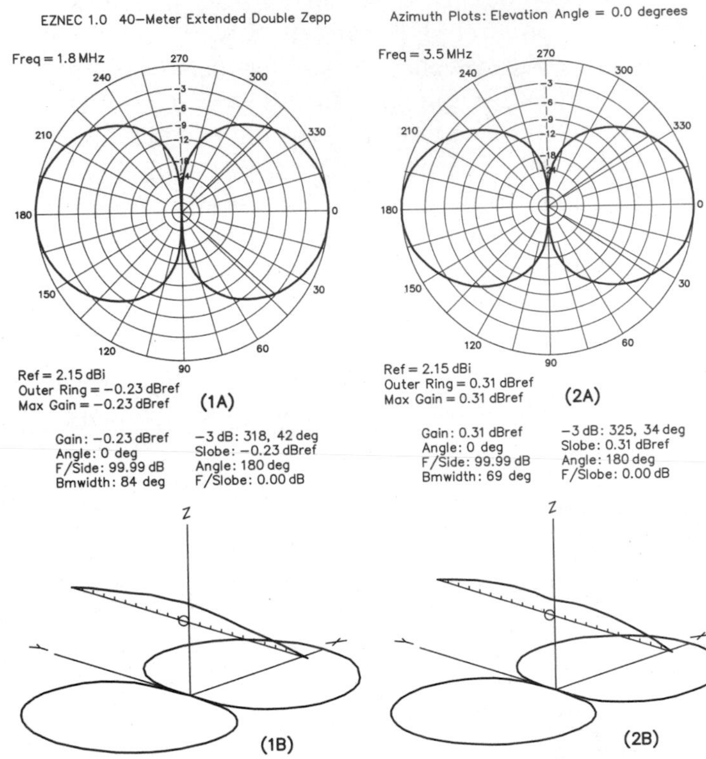

Figure 6-9 — These drawings illustrate the performance of a 40 m extended double Zepp antenna when it is used on the 160, 80, 40, 30 and 20 m bands. Part A of each plot shows the antenna azimuthal pattern for that band.

Plots 1 and 2 show that 160 and 80-m performance are pretty much alike. In both instances the free-space pattern looks like a classic dipole antenna: a figure 8 broadside to the wire. Gain on 80 m is only 0.31 dBd while there is a loss of 0.23 dBd on 160 m.

On 40 m, the EZNEC plot for this 40-m extended double Zepp (EDZ) mirrors the pattern shown in The ARRL Antenna Book. *The two major lobes have become very pronounced and four smaller lobes have appeared. Gain is 2.8 dBd on 7 MHz. The pattern is broadside to the antenna wire. On-the-air performance is much better than my old 80 m double Zepp.*

The 30-m plot shows a much more skewed radiation pattern, with four major lobes. This EDZ exhibits a 1.46 dBd gain on 30 m, but you have to be careful when erecting the antenna to make sure the lobes are positioned properly to enhance reception from the areas in which you are interested.

The 20-m performance of this antenna is a little more bizarre. The four major lobes have become more enhanced and six minor lobes have appeared at right angles to the wire. Gain (as measured on the major lobes) is almost 3 dBd, but care must be taken when erecting this antenna to ensure proper positioning of the lobes.

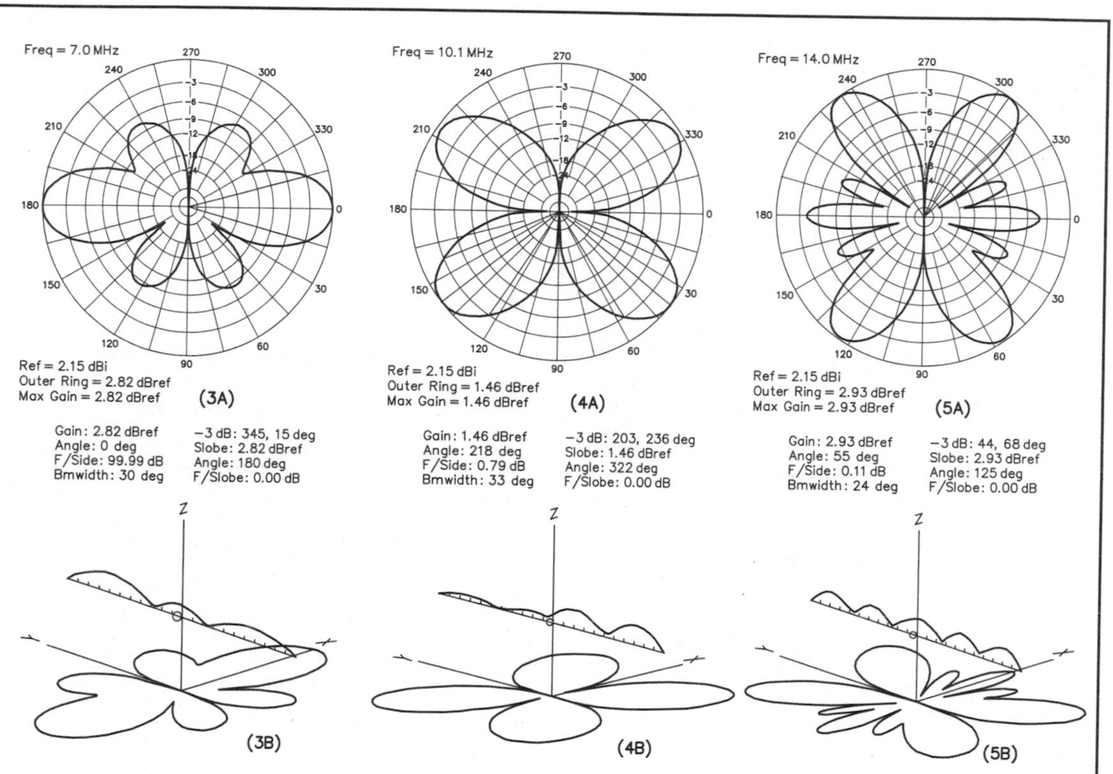

THE G5RV

In 1982 I had the delightful experience of being able to meet and talk at length with Louis Varney, G5RV, the man whose antenna bears his call sign. I met Lou at the RSGB HF Convention at Oxford University while stationed in England. The G5RV had become a very popular antenna in the UK, and had started to pick up a following in the US as well. With an overall length of only 102 feet end-to-end (51 feet per leg) and fed with 300 Ω twin lead terminated into 72 Ω coax, the G5RV offers good multiband performance. **Figure 6-10** gives typical construction details. Originally, Lou designed this antenna for 20 m, so it did offer some gain on this band. With the use of an antenna tuner, however, you can operate 80 through 10 m with this antenna and radiate a reasonably good signal to boot.

The key to making the G5RV antenna perform in a multiband

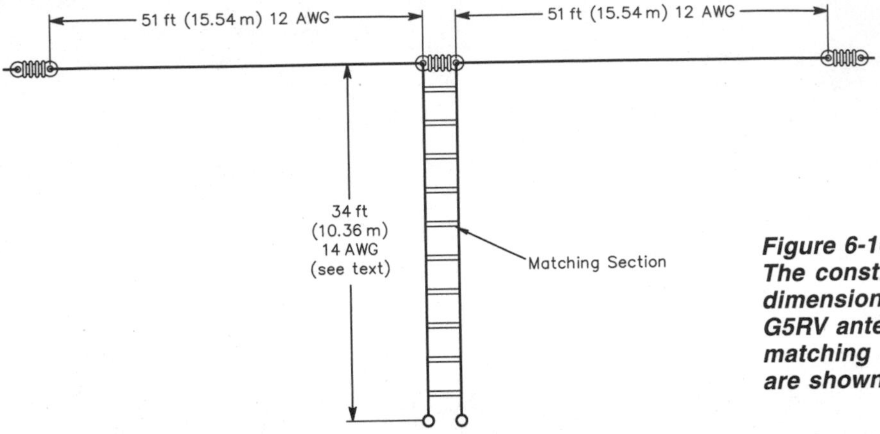

Figure 6-10 — The construction dimensions of the G5RV antenna and matching section are shown here.

role is the antenna tuner. Many try using it without a tuner and suffer disappointing results. A tuner tames the wild impedance excursions, properly matching the radio and the antenna on 80 through 10 m. To have a shortened 80 m antenna that still performs well is a godsend to many people who don't have the room to erect a full-size 80 m dipole.

WINDOM ANTENNAS

The Windom antenna was first described in the September 1929 issue of *QST* by Loren G. Windom, W8GZ. This antenna was fed with a single wire connected about 14% off center, and was operated against an Earth ground. Some amateurs have adopted this antenna for use with coaxial cable feed line. The coax-fed antenna is called an off-center-fed (OCF) dipole. There is really no space savings associated with this OCF antenna, but you do get some directivity. Mismatching the feed point causes RF currents to be induced in the feed line which, when properly controlled by a coaxial RF choke, yields both vertically and horizontally polarized radiation from one antenna. This is great for the operator who wants both close and long haul contacts. Currently the Carolina Windom marketed by The Radio Works, is the only commercially available offering of this antenna. I have used several versions of the Carolina Windom with excellent results on all bands, 160 through 10 m. This antenna requires a matching network for all-band operation.

CAROLINA WINDOM

80, 40, 30, 20, 17, 15, 12, 10 Meters + SWL

GENERAL MOUNTING REQUIREMENTS

*Mounting height for vertical section: >30'
*Minimum angle between legs = 120 degrees
*Minimum height at ends = 8'

GROUND SPACE NEEDED

Configuration vs. Length needed

Flat top:	134'
Inverted-V @ 40' height:	77' + 40' = 117'
Inverted-V @ 60' height:	73' + 34' = 107'
Inverted-U @ >30' height:	length = 114'
Inverted-U Bend =	10' short leg; 15' long leg
Sloper @ 40' =	128'
Sloper @ 60' =	121'

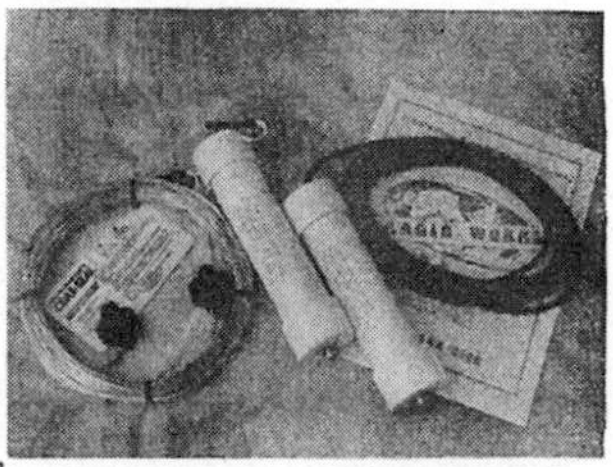

Recommended mounting configuration:

Flat-top, suspended between two tall trees located >140' apart. Antenna may be supported as an Inverted-V. Bending the elements will alter the radiation pattern on some bands.

CAROLINA WINDOM

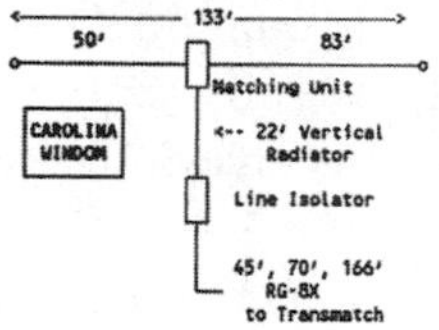

We are continually improving our products, specifications are subject to change.

Optimum transmission line length: 45', 70', 166. Other lengths may be used. 1/2 λ lines should be avoided.

SPECIFICATIONS

Freq. coverage:	80 - 10 meters
Gain:	As much as 10 dbd*
Radiator length:	Horizontal 132' Vertical 22'
Polarization:	Both vertical and horizontal components
Feed line:	50 ohm Coaxial cable
Matching method:	Matching Transformer + User's transmatch
Transmatch needed:	Yes on 40 - 10 M, recommended on 80
SWR:	Low, adj. w/ transmatch
Power Rating:	1500 Watts
Recommended Ht:	>35'- Usable at 30'
Radials?	Not required

* Based on user reports, field evaluations, and product reviews.

The Radio Works manufactures the Carolina Windom antenna, which is a variation on the true Windom antenna. Rather than bringing the single-wire feed line directly into the shack, the Carolina Windom uses a line isolator to connect a coaxial feed line, which then comes into the shack.

WIRE BEAMS

Sooner or later, everyone wants some kind of directional antenna. If you have the money, time and real estate available, it is quite possible to erect a nice tower and rotatable beam antenna that will be a great DX-getter. There are ways to provide directivity and gain at reasonable cost and still keep your feet planted firmly on terra firma, however.

Wire beams are not a new idea. John Kraus, W8JK, worked wonders with his 8JK directional antenna in the 1930s. **Figure 6-11**

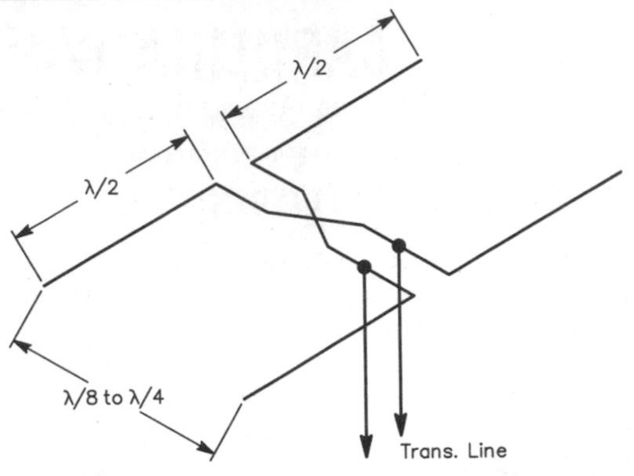

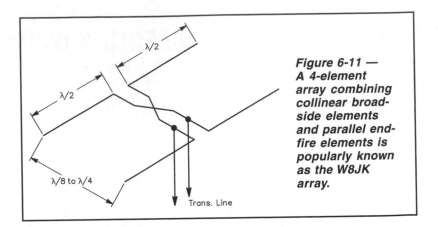

Figure 6-11 — A 4-element array combining collinear broadside elements and parallel end-fire elements is popularly known as the W8JK array.

shows the basic configuration of this antenna.

I was treated to quite a sight when I journeyed west in 1996 to attend the ARRL Field Day with the premier QRP FD group, The Zuni Loop Mountain QRP Expeditionary Force (aka: The Zunis). From their 7600 foot (above sea level) location in the San Gabriel Mountains overlooking Los Angeles, I saw them erect their "secret weapon" for 40 m: a 4-element full-size 40 m wire beam antenna! This antenna was fix pointed east and hung between a bunch of trees at 60 feet above ground. To say that the sight of this antenna was awesome would be an understatement. No wonder the Zuni's signal is so potent on the East Coast.

Fred Turpin, K6MDJ, and Cam Hartford, N6GA, two of the Zuni's antenna engineers, told me that over the years they had experimented with larger arrays, but decided to use just 4 elements because their signal tended to miss the East Coast when more than 4 elements were present! As Field Day progressed, they would lower the non-driven elements one at a time, to alter their coverage area on 40 m to get more contacts.

Paul Stroud, AA4XX, near Raleigh, North Carolina, is another wire beam aficionado. Paul's 3-element wire beam, which points north, has netted him two world records for miles-per-watt on 40 m.

There is nothing wrong with using a fixed-direction wire beam to work DX, either. One 3-element wire beam pointed in the general direction of Europe and another one pointed over the pole toward the Pacific Rim, would be great DX antennas.

If you have the physical space and trees in the right places, wire beams may be just what you need to work the world with a couple of watts. Check out *The ARRL Antenna Book* for dimensions and feed arrangements. Wire beams are inexpensive (when compared to rotatable arrays) and you don't have to worry about rotators, extra cabling or a tower structure.

LOOPS — DIRECTIVITY AND LOW NOISE

Loop antennas, whether we are talking about Delta Loops or Squares or some other geometric layout, are a great way to pick up some directivity and achieve low-noise reception. The loop is a "closed circuit" type of antenna that tends to reduce noise. The loop, depending on its size, either $1/2\ \lambda$ or $1\ \lambda$, can be directive either 90° to the plane of the wire or off the ends of the wire. **Figure 6-12** shows a simple installation. This comes in handy when

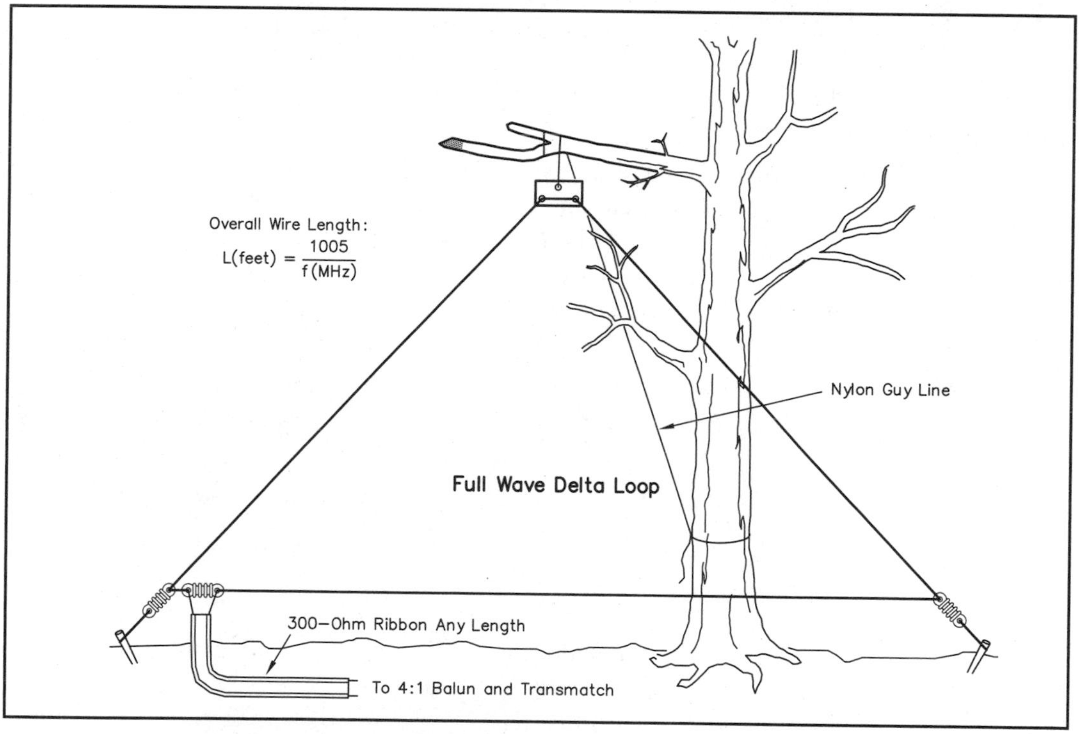

Figure 6-12 — Here is an example of a delta loop antenna that only requires one support structure. Cut for the 40 m band, this antenna can be used on all bands, 40 through 10 m. It works especially well on bands that are not harmonically related to the fundamental frequency, such as 30, 17 and 12 m. The greater the antenna height, the better the DX performance. Feeding the antenna at the apex rather than at a lower corner increases effective height. Maximum radiation is broadside to the loop.

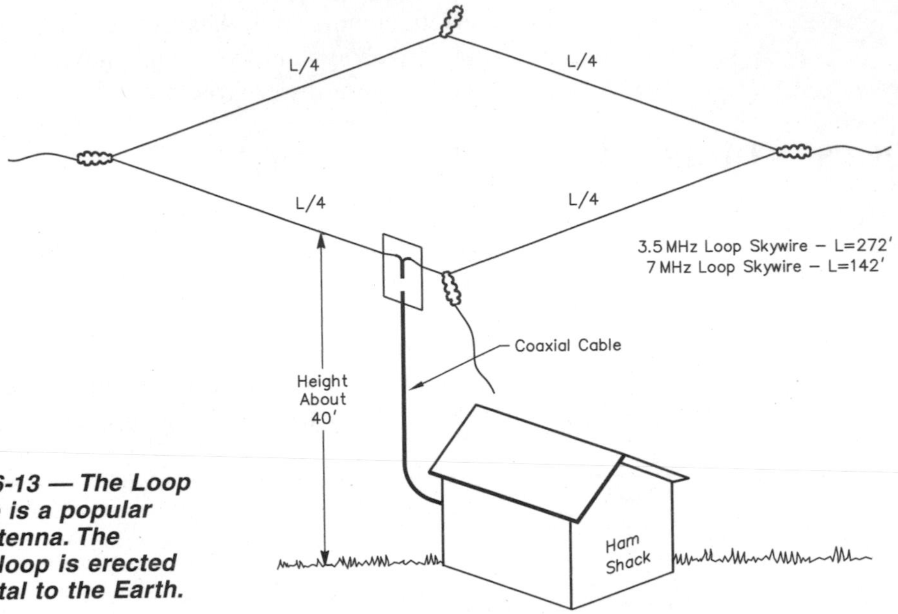

Figure 6-13 — The Loop Skywire is a popular loop antenna. The square loop is erected horizontal to the Earth.

trying to minimize signals or interference from a specific direction. Ernie, W8MVN, uses a pair of 40 m loops suspended off his tower to radiate one of the most potent 40 m QRP signals in the US!

The formula for figuring the dimensions of a 1 λ loop antenna is: 1005/MHz. This will yield a full size driven element for whatever band you desire. **Figure 6-13** depicts a multiband, horizontal 1 λ loop that is capable of working on 80, 40, 20, and 10 m. The cost is minimal since this antenna uses only wire, ceramic insulators and coaxial cable. Using a Transmatch will let you tune this antenna on the 30, 17, 15 and 12 m bands as well. Erect this antenna as high as possible using trees or masts.

ANY DAY YOU'RE VERTICAL IS A GOOD DAY!

When you need an antenna with a small "footprint" that is capable of producing DX contacts, then a vertical antenna is a good choice. Verticals come in several varieties: $^1/_4$ λ trapped verticals and linear loaded verticals. For many years $^1/_4$ λ trapped verticals dominated the market. The Hy-Gain 14AVQ, 18AVQ, Newtronics

Figure 6-14 — The Cushcraft AV-5 is one example of a commercially manufactured trap vertical antenna.

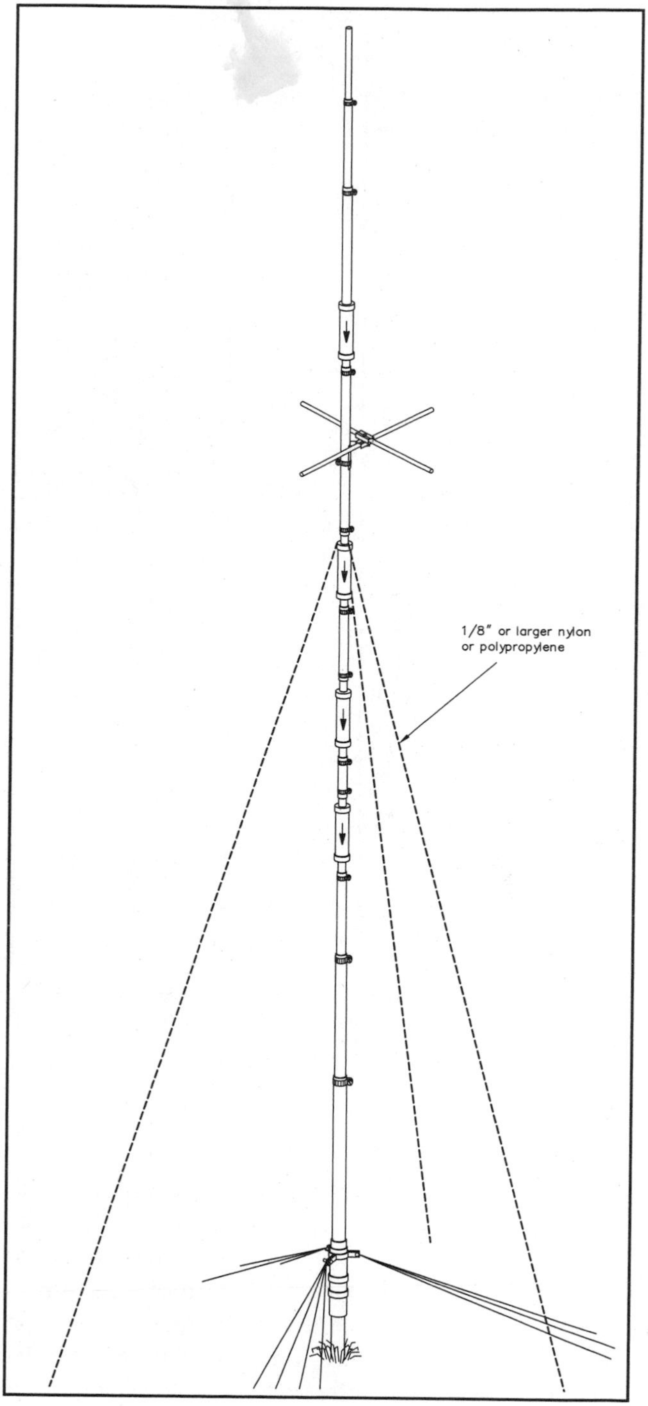

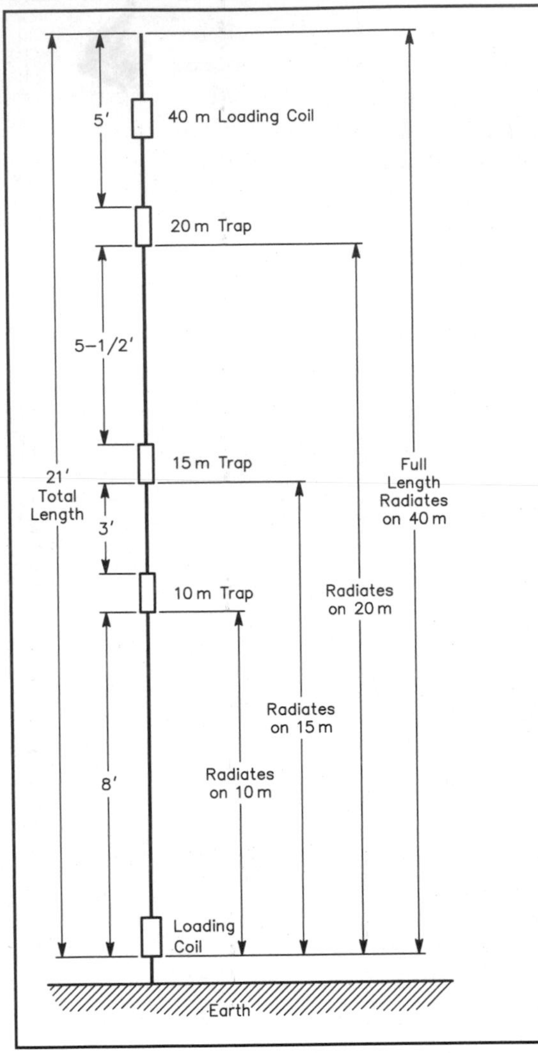

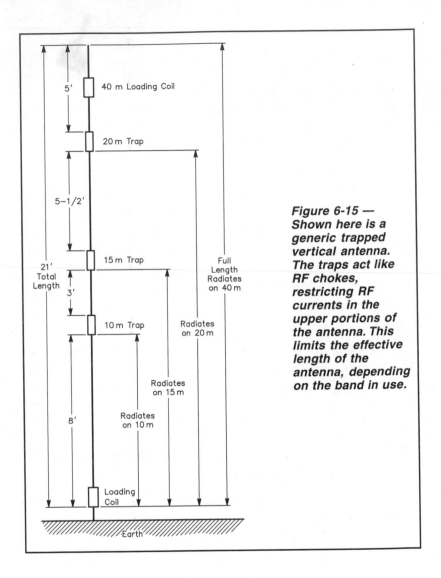

Figure 6-15 — Shown here is a generic trapped vertical antenna. The traps act like RF chokes, restricting RF currents in the upper portions of the antenna. This limits the effective length of the antenna, depending on the band in use.

4BTV, 5BTV, and the Cushcraft ABV-5 are examples of $^1/_4$ λ trapped vertical antennas. See **Figure 6-14**. Each portion of the antenna is separated by a parallel tank circuit — called a trap — that acts as an RF choke, keeping the RF restricted to only the proper portion of the antenna. As the frequency of operation decreases, the traps act as a short and electrically connect the next section of the antenna. **Figure 6-15** shows a basic $^1/_4$ λ trapped vertical antenna. The 10 m section is the first section above the feed point. The 10 m trap acts as an RF choke at 28 MHz and prevents the RF energy from proceeding up the antenna mast to the 15 m section. When the operating frequency is reduced to 21 MHz, the 10 m trap presents a very low impedance to the RF energy, allowing it to flow into the 15 m section. Likewise, when the frequency is dropped to 14 MHz, the 10 and 15 m traps present a low impedance to the RF energy, allowing the RF to flow into the 20 m section of the antenna.

Quarter-wave trapped verticals are fine antennas but they are not the most efficient vertical antennas available. Traps are lossy and can go bad, especially if they are subjected to very high power levels. Figure 6-15 shows the basic principle of how a trap vertical antenna operates.

Linear loaded verticals like the Butternut HF6V or HF9V,

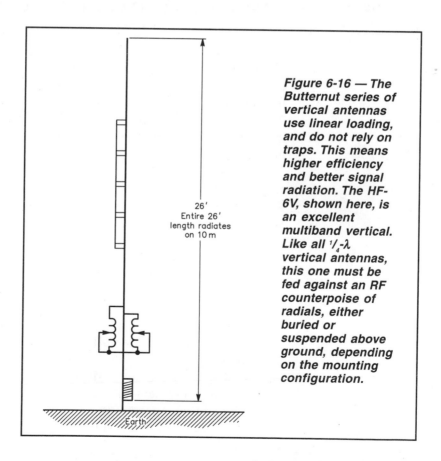

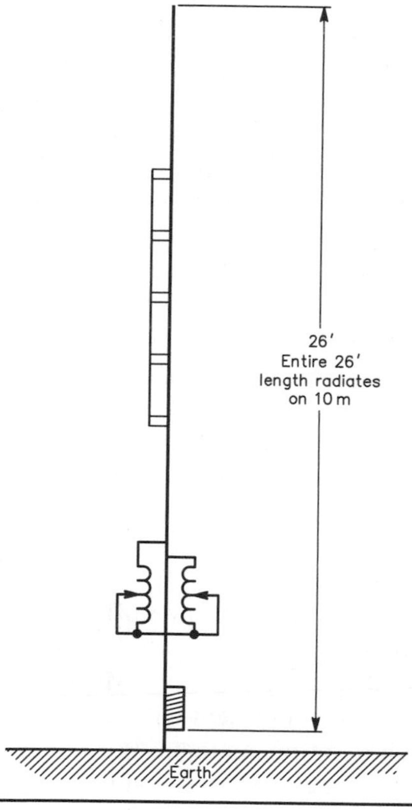

Figure 6-16 — The Butternut series of vertical antennas use linear loading, and do not rely on traps. This means higher efficiency and better signal radiation. The HF-6V, shown here, is an excellent multiband vertical. Like all $1/4$-λ vertical antennas, this one must be fed against an RF counterpoise of radials, either buried or suspended above ground, depending on the mounting configuration.

and the Cushcraft R-5, R-7 and R-7000 are more efficient but they are also slightly longer. Instead of traps these antennas use inductive circuits and linear loading stubs to tune the antenna on each band. The entire length of the antenna radiates on each band. See **Figure 6-16**.

Vertical antennas are very good DX antennas because, when properly installed, they have a very low angle of radiation. A low angle of radiation is desirable when working DX because the lower the take-off angle of the RF wave from the antenna, the farther it travels before encountering the refractive properties of the ionosphere. This means that the skip distance between the antenna and where the RF wave returns to the earth is longer. The longer the skip distance the farther you can talk. Ergo, low radiation angles are great for working DX.

THE PROS AND CONS OF ERECTING A TOWER AND BEAM ANTENNA

There is a great controversy between the inhabitants of one of the QRP reflectors about whether or not using a beam antenna is against the spirit of QRP. I'm not exactly sure what spawned this ongoing debate between the factions, but it certainly is silly. The spirit of QRP dictates that we use all methods available to make our stations more efficient. Beam antennas certainly fall under that heading.

Should you desire to erect a beam antenna and associated tower and rotator, be prepared to do a lot of homework and lay some groundwork in the neighborhood and at City Hall before you even contemplate starting construction.

There is an old saying I learned while serving in the US Air Force: "It's easier to beg forgiveness than get permission." This applies across the board to many situations in life. Dealing with City Hall about a tower installation is *not* one of them! They can and will make you take down your newly erected antenna should they not be "in the loop" from the getgo. Take the time to check with the folks in the zoning office, city engineer's office and the planning commission at City Hall just to be on the safe side. Get any agreements in writing. Most of the time these people are grossly ignorant of ham radio and associated antennas. It will be your job to enlighten them. For this task, you must be prepared.

The same goes with your neighbors. While you can be

stealthy and hide wire antennas in the trees and under the eaves of your house or apartment, a tower is sort of a dead giveaway that you are serious about the ham radio hobby. Be prepared to explain, re-explain and re-re-explain what ham radio is all about, what you do with all your radios, and why you want to ugly up "their" neighborhood with a steel and aluminum structure that looks like something out of Star Wars.

The ARRL has a complete antenna zoning package that includes the FCC PRB-1 ruling and other information. Don't be afraid to ask. Get the package, study it and become prepared to explain why you want to erect this monstrosity. Believe me, you are in for an uphill battle. With the planned communities and their Draconian covenants and stipulations, you will need all the help and preparation you can get prior to taking on the neighbors and the local city government.

One thing you might think about is a fall-back position, using a smaller tower and beam antenna than you actually want to install. Don't forget, in trying to pitch this idea it never hurts to negotiate to get what you want. Let's say your original plan calls for a 90-foot guyed tower and a KT-34XA Yagi. The neighbors are furious and they say; "Not no, but *hell* no!" Try negotiating with them. Find out what they don't like about the installation. Pin them down. Make them accountable for their reasoning. Just saying "No" is not good enough. Then re-propose something smaller, like a 48-foot unguyed crank-up tower and a smaller Yagi antenna, like a TH7DX. By trying to work with the neighbors and showing your willingness to scale down your antenna system to keep peace in the neighborhood, you have demonstrated your concern for their feelings and a desire to negotiate in order to satisfy both your and your neighbor's needs.

Once you get permission and all paperwork is completed at City Hall (don't be surprised if the city requires you to obtain a zoning variance and furnish engineering drawings to the city engineer's office), it is time to start erecting the tower. Here's when you find out who your friends really are! Follow the manufacturer's instructions on the dimensions and amount of concrete for the base of the antenna. Carefully follow all the instructions for a safe tower erection. Use hard hats, gloves and other safety equipment to ensure that no one gets hurt. It is a very good idea at this point to ensure that several members of your antenna crew are experienced in erecting towers and antennas.

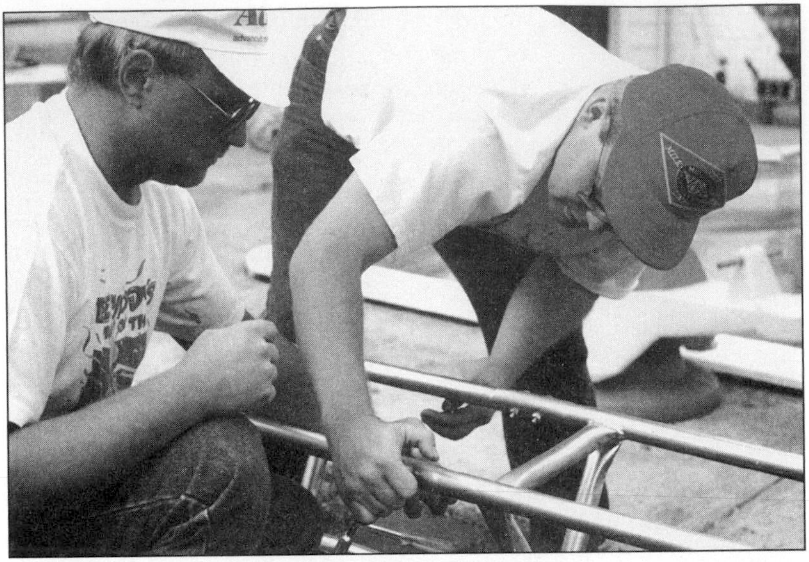

It's antenna party time at K7SZ! This picture shows Fran Slavinski, KA3WTF, and Dave Carey, N3PBV, checking the bolts on a section of tower prior to erecting it against the side of the house. Check, recheck and re-recheck is the name of the game when it comes to a tower and beam installation. **(K7SZ Photo)**

Nothing beats experience.

Once the concrete has cured and the tower is installed, you now have to install the rotator and antenna at the top of the tower. If you have the use of a bucket truck (cherry picker) then the task is simplified. Most of us use the tried and true method of belting oneself off the side of the tower and hoisting the antenna into place using ropes and a gin pole, however. This is hard and potentially dangerous work. Again, experience counts, so be sure that the crew has some seasoned veterans on board.

As with any antenna system, feed the antenna with good-quality coaxial cable. At K7SZ, I use RG-213 for the HF runs and Belden 9913 for the VHF/UHF antenna runs. Quality coaxial cable ensures that your QRP signals are making it to the antenna with minimal losses. Be prepared to replace the coax every 3 to 5 years, depending on your climatic environment. Ultraviolet rays along with severe cold or hot temperatures degrades coaxial cable. Play it safe and change out your coaxial runs on a regular basis.

A tower and beam installation is a financially and emotionally intense time. A lot of work goes into getting something big up into the air. The rewards are worth every sleepless night and spent penny. Using a high performance HF beam antenna has distinct advantages. For one thing you can concentrate your QRP signal in a given direction. At the same time, you limit the signals coming in from the back and sides of the antenna, so your signal-to-noise ratio improves. This type of efficiency is definitely in keeping with the spirit of QRP!

THE QUEST FOR CONDO EQUALITY

It is a fact of modern day life that many of us who wish to pursue a ham radio hobby must contend with the restrictions of living in a condo, planned community or townhouse arrangement that either forbids or greatly impedes your ability to erect external antennas. Unfortunately, there are no easy solutions.

Jim Kearman, KR1S, in his book *Low Profile Amateur Radio* describes how he used indoor loop antennas with good results when he moved into an apartment. If you live in an apartment, condo, townhouse or planned community, Jim's book is a must read. Browse the hamfest tables and other sources for a used copy of this book, because it is no longer in print.

Dave Carey, N3PBV, lives across the street from me in south

***Fran , K3BX, and Dave, N3PBV, are aligning the elements on the TH7 while it sits on a pair of saw horses in the backyard. Checking and rechecking all the mounting hardware is a must before hoisting a large, high performance beam into the air.* (K7SZ Photo)**

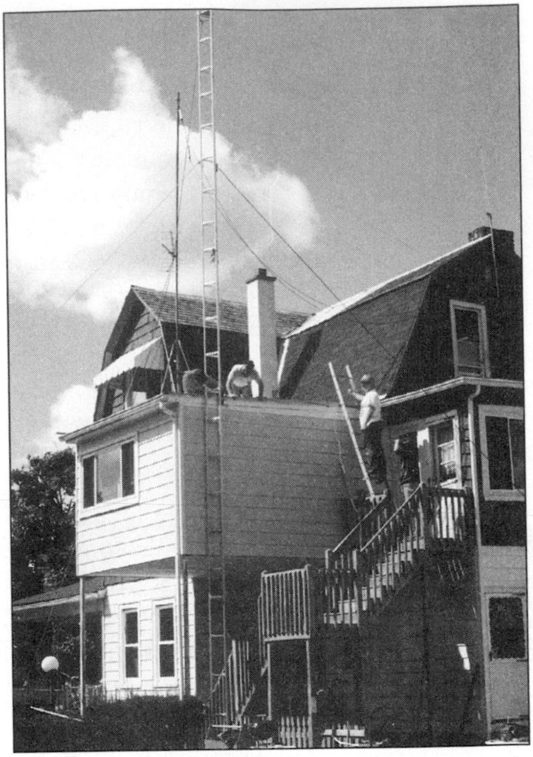

***It's up! The antenna crew walked the tower up from the ground level while the author (with the help of a Federal Express driver) pulled on a rope from the other side of the dining room extension. Once in place, the two tower legs that rest against the wall of the dining room were secured with long bolts through the framework of the building. Guys were added later just below the rotator plate to fully stabilize the installation.* (K7SZ Photo)**

Wilkes-Barre, Pennsylvania. Dave and his wife, Jenny, are both licensed hams. Dave wanted to erect some simple antennas but Jenny was reluctant. She wanted to keep the aesthetics of the house and yard pristine. What to do, what to do?

Dave solved the problem by placing dipole wires beneath the soffits of his roof when he reroofed the house! He first modeled the wire installation using W7EL's *ELNEC* antenna modeling software and found that they would play pretty well on 40 to 10 m. The next step was to lay the wire down, after the roof shingles had been removed. The new roof went down over the wires and Dave has a nice little invisible antenna farm that keeps Jenny happy.

Do Dave's antennas work? You betchum, Red Rider! Since Dave is a QRPer (I wouldn't let him get a QRO rig!) he gets on the air using a NorCal 40 and his hidden antenna system with good results. One point to remember: since the antenna wires are in proximity to the wooden structure of the house, at power lev-

***Here Mike Ratican, W9RAT, tightens the boom-to-mast bolts on the TH7. Between Mike and my son Jamie, all the tower work was completed in short order. Although not clearly visible in this picture, Mike is wearing a lineman's climbing belt with climbing lanyard, and he is belted off so he can work with both hands. As with all antenna work THINK SAFETY and use the proper safety equipment.* (K7SZ Photo)**

els of 100 W or higher, there is a danger of fire, since the ends of these wires are high impedance points on the antenna and will develop very high RF voltages. With 5 W or less this isn't a problem — another reason for QRP!

Another trick to make your antenna invisible is to load up the rain gutters. Since most rain gutters are one-piece aluminum, they provide a long run of conductive material that can be loaded like an end-fed wire via an L-Match tuner. See **Figure 6-17**. Loading the rain gutters works as long as the dwelling is either wood or sided with vinyl siding. Aluminum sided structures skew the radiation patterns of this type of antenna system. Be careful of any downspouts, since these may be at a voltage node (high impedance point) on the antenna and could result in a nasty RF burn to an unsuspecting person.

I have loaded rain gutters with good results using nothing more than a piece of #18 or 20 insulated wire with an alligator clip on one end (to clip to the rain gutter) and the other end termi-

Figure 6-17 — Rain gutters and TV antenna installations can be used as inconspicuous Amateur Radio antennas.

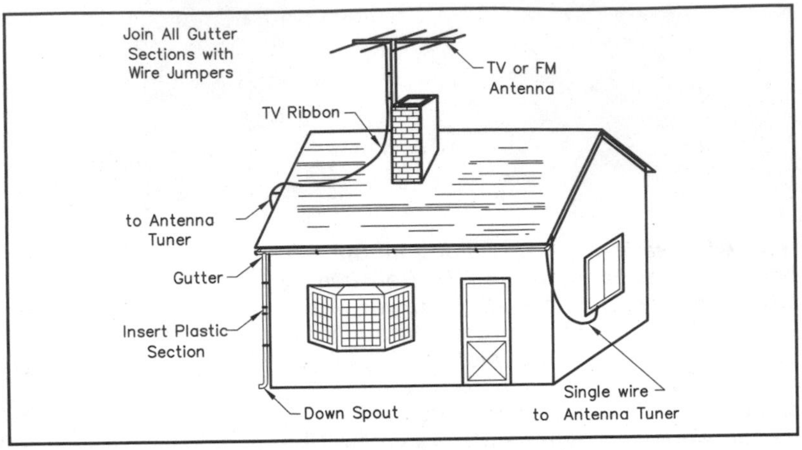

The ARRL PRB-1 Package, Zoning Restrictions and the Radio Amateur

Amateurs have been plagued by zoning restrictions for many years, but this became a greater problem during the '90s as local governments responded to the need for cellular and PCS towers. Because these towers were previously unregulated, some local governments enacted new ordinances to deal with these commercial towers. In the process, they sometimes over regulate Amateur Radio antennas. Most communities have some sort of building permit requirements for amateur towers ("antenna support structures") and PRB-1 doesn't eliminate those permit requirements. PRB-1 does, however, explain to local planners why amateurs need antennas despite what irate neighbors may say!

PRB-1 is the amateur defense to overly restrictive local antenna ordinances. (PRB stands for the FCC Private Radio Bureau. This was the FCC Bureau that handled Amateur Radio matters at the time PRB-1 was released in 1985. It was renamed the Wireless Telecommunications Bureau in the early '90s.) PRB-1 is the partial preemption of local zoning ordinances for amateurs. The PRB-1 document states that an amateur antenna structure may be erected at heights and dimensions sufficient to accommodate Amateur Service communications. State and local regulation of a station antenna structure must not preclude Amateur Service communications. Rather, it must reasonably accommodate such communications and must constitute the minimum practicable regulation to accomplish the state and local authority's legitimate purpose of protecting the health and safety concerns of community citizens. The 11 page PRB-1 document can be found on the ARRL Web at: **www.arrl.org/field/regulations/**. Information on the ARRL Volunteer Counsel Program also appears at this web site. This program provides the names, addresses and phone numbers of hams who are lawyers and who have agreed to provide a free initial consultation with ARRL Members to determine if legal services may be necessary. Contact information for ARRL Volunteer nated in a banana plug that fit into the center of the SO-239 connector on the back of the tuner. A couple of RF counterpoise wires were added to the ground side of the tuner and this setup played well when I was in Altoona, Pennsylvania during a QRP contest weekend. Don't forget to scrape away some of the paint on the aluminum gutter so the alligator clip makes good electrical contact.

Apartment dwellers who live higher than a second floor, can use the old end-fed-wire-out-the-window trick, as discussed earlier in this chapter. Fed against a counterpoise inside the apartment, this antenna works like a top-fed vertical antenna. Not the greatest, but it does work.

Indoor loop antennas that are strung around the ceiling or

Consulting Engineers is also included on the web site. Volunteer Consulting Engineers may be able to help you decide what type of engineering services you may need to meet antenna zoning ordinances. The ARRL Antenna Height and Communications Effectiveness study appears at this site. Specifically intended to be part of a presentation to local zoning officials, it demonstrates that amateur antennas at 75 feet are more effective than antennas at 35 feet.

In many instances, amateurs need additional material than what appears on the ARRL Web page. PRB-1 doesn't specify what is meant by "reasonable." The *ARRL PRB-1 Package* can be very useful, since it includes over 200 pages of material that can be part of a presentation before local zoning boards. In addition to what appears on the ARRL Web, the package includes several Federal District Court cases that show that "reasonable" means 65 or 75 feet in height. The package includes FCC letters demonstrating the FCC's sole jurisdiction over interference, sample ordinances, *QST* articles on antenna regulation and other helpful information. The cost of the package is $10 for ARRL members ($15 for non-members), which helps offset the cost of reproducing and mailing the package.

It is important to note that PRB-1 does not apply in situations such as covenants, deed restrictions and leases. PRB-1 was not meant to cover such agreements, which were entered into *voluntarily* by individuals. In such cases, working with local homeowners' associations or landlords by "playing up" the public relations value of Amateur Radio is an option. A lawyer may be able to help explore other options.

Amateurs who wish to order the ARRL PRB-1 package or those with questions about zoning related concerns may contact the Regulatory Information Branch at ARRL HQ by letter at 225 Main St, Newington, CT 06111; by e-mail at **reginfo@arrl.org**; by fax at 860-594-0259; or by phone at 860-594-0236. — *John C. Hennessee, N1KB, ARRL Regulatory Information Specialist* behind draperies that take in the entire dimensions of a room are a good bet. **Figure 6-18** shows one way to support the wire along the ceiling. Treat this antenna just like a regular loop antenna and feed it with transmission-quality 300 Ω twin lead via a tuner with balun. While they won't work as well as a full-size exterior loop, these small loop antennas will allow you to get a signal on the air in good fashion. **Figure 6-19** shows one way to incorporate the matching system into the antenna feed point. The $^1/_4$-λ counterpoise helps control any stray RF that might appear on the outside of the coax shield.

During my stay in base housing at RAF Mildenhall, UK, I was hard pressed to erect any outside antennas. My house did have an attic/crawl space above the living area. This was perfect for installing a half-size G5RV dipole antenna. I placed the apex of the dipole near the very top of the rafters at one end of the attic. Using #14 insulated wire as elements, I ran each dipole leg down the end rafters to a point near the attic floor, using a staple gun to secure the wire as I went. I stopped when the dipole legs got about 12 inches from the attic floor. From here I ran each leg perpendicular to the rafters (parallel to the floor) until I ran out of wire. Placing an end insulator on the end of each dipole leg, I then secured the ends to a nearby rafter with a length of small rope. This antenna was fed using 450 Ω open-wire line, which was routed down into my shack through two very small holes in the ceiling of the shack, directly beneath the antenna. Using this antenna I worked 85 countries in about one year of casual QRP operation using a Ten-Tec Argonaut 515. And *no one* in the housing area knew I was on the air except me and my immediate family!

In *The ARRL Antenna Compendium, Volume 2*, Kirk Kleinschmidt, NTØZ, described a wire tri-band Yagi beam that he installed in his attic. **Figure 6-20** shows the basic construction details to give you an idea of what can be done.

Attics are great places to put up trapped dipole antennas. Cushcraft makes a series of rotatable dipoles that work very well suspended in an attic and fed with coaxial cable.

Finally, let's explore "invisible" antennas. By invisible I mean that they are nearly impossible to spot by the casual observer on the ground. The wire can be bare or covered. If you decide to use covered wire, pick a color that is neutral, like gray or light blue, as it will blend in with the sky very nicely. Using very small diameter (#24 to #30) wire, it is possible to erect a nearly invisible end-

Figure 6-18 — An indoor wire antenna can be supported with self-stick cup hooks. Placed high on the wall, most people will never notice them. If you think someone might spot them and ask embarrassing questions, you can always hang a decoration from them!

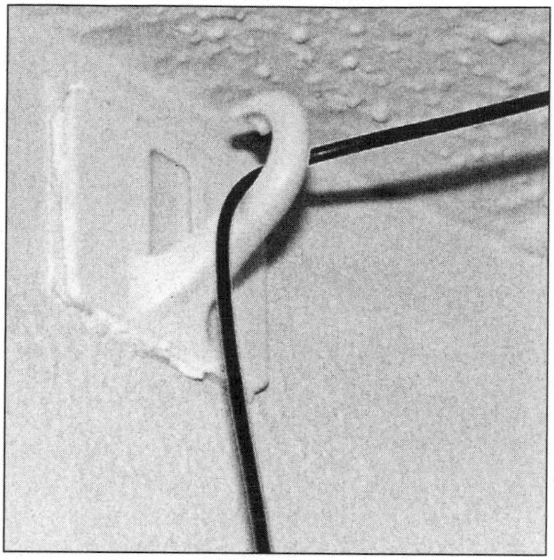

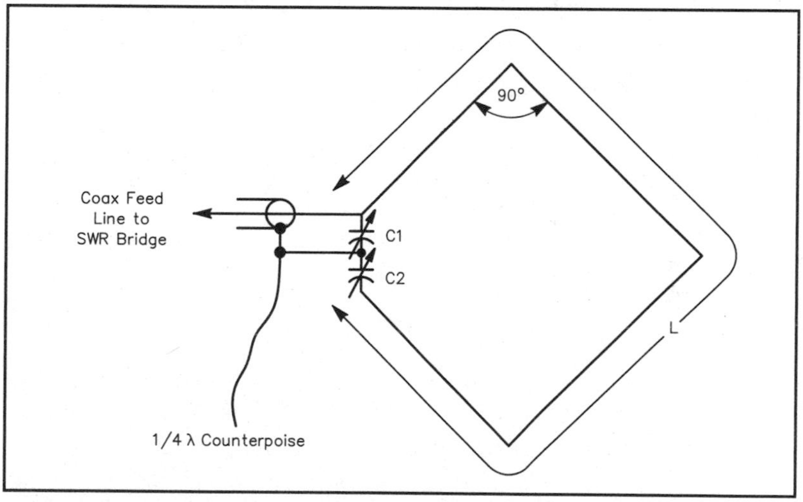

Figure 6-19 — Here is a simple matching system for a wire loop, as used by Rod Newkirk, W9BRD. For QRP operation, receiver-type capacitors will suffice. C1 is about 300 pF and C2 is about 100 pF. While this drawing shows the loop supported in a diamond shape, use whatever configuration suits your location. Loop diameter should be slightly less than $^1/_4$ λ, or about 25 feet for 40 m. Because this antenna is fed with an unbalanced feed line, the addition of a $^1/_4$-λ counterpoise may be necessary. The counterpoise may be run around the edge of the room or under the carpet. As long as you only operate QRP, you should have no trouble with high RF voltages along the wire.

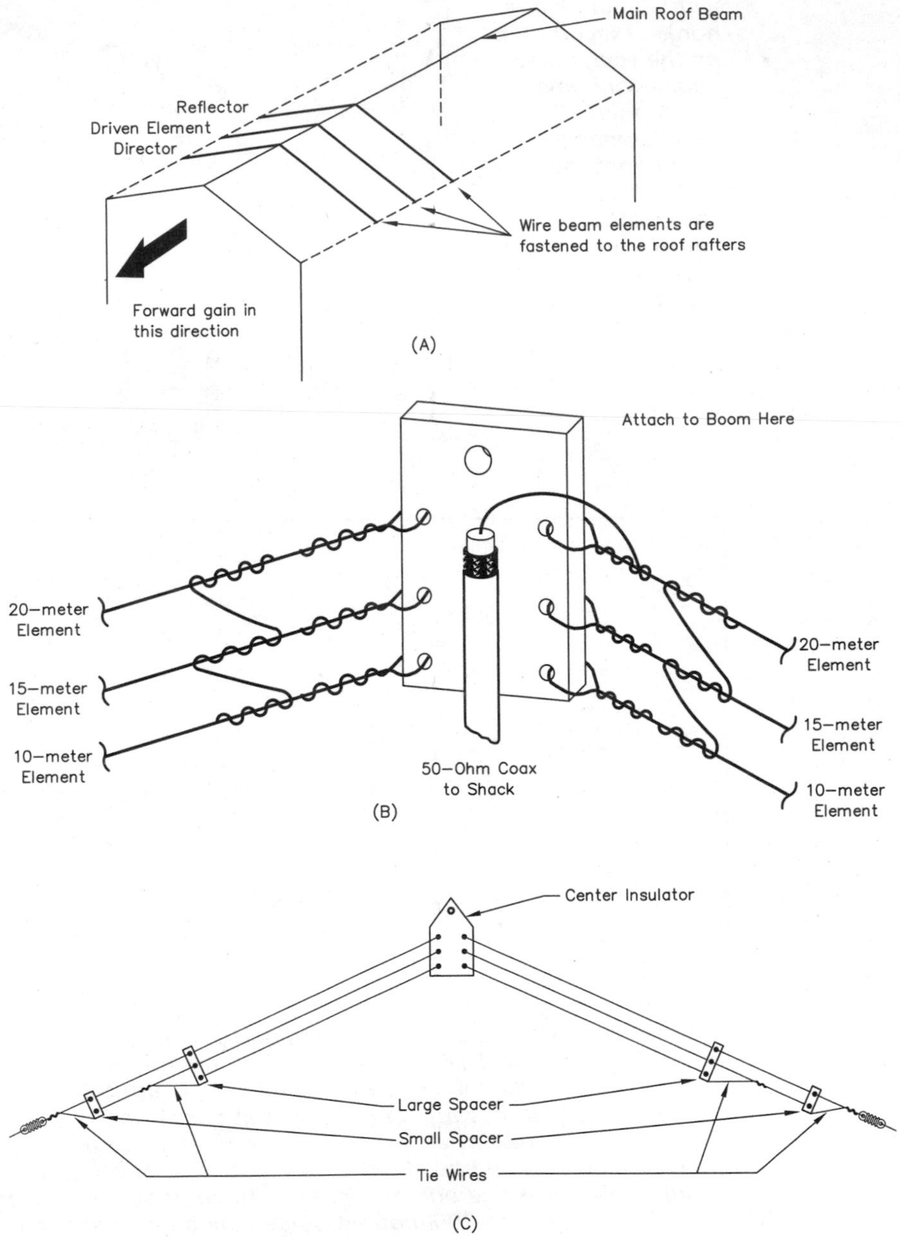
Main Roof Beam
Reflector
Driven Element
Director
Wire beam elements are
fastened to the roof rafters
Forward gain in
this direction
(A)
Attach to Boom Here
20–meter
Element
15–meter
Element
10–meter
Element
50–Ohm Coax
to Shack
20–meter
Element
15–meter
Element
10–meter
Element
(B)
Center Insulator
Large Spacer
Small Spacer
Tie Wires
(C)

Figure 6-20 — You can take the idea of an attic dipole one step further and build a tri-band beam, like Kirk Kleinschmidt, NTØZ, did. Kirk described his antenna in **The ARRL Antenna Compendium, Volume 2.** *Part A shows how the wires are suspended from the rafters, along the roof line. Part B shows how the feed line connects to the driven elements at the feed point. The center supporting insulator is made from a 2 × 3-inch piece of plastic or acrylic sheet, such as Plexiglas. Bare copper wires tie the driven element halves together for each band. Part C shows the completed beam element, including placement of plastic spreaders and center insulator. The tie wires are connected to the bottom of the spaces, and not to the ends of the shorter elements. You can add reflector and director elements to the driven element. Kirk used a rope "boom" strung along the peak of the attic to hang the beam elements.*

RF Exposure and Indoor Antennas

The FCC Rules establish maximum permissible human exposure (MPE) limits to the electric and magnetic fields as well as the power density produced by the RF energy from any transmitter. In Part 97 the Rules also say

§97.13 Restrictions on station locations.

(c) Before causing or allowing an amateur station to transmit from any place where the operation of the station could cause human exposure to RF electromagnetic field levels in excess of those allowed under §1.1310 of this chapter, the licensee is required to take certain actions

One of those actions is to perform a routine RF environmental evaluation, as described in §1.1307 of the Rules, if the station power exceeds certain limits. QRPers will always be using power levels below the limits that establish when such an evaluation *must* be carried out. They must still ensure that no one will be exposed to more than the MPE limits, however.

Indoor antennas present one situation in which it is possible that even a QRP transmitter could produce enough RF energy to exceed those limits. A dipole antenna in the attic above a bedroom or strung along the wall next to your neighbor's apartment, for example, could result in more than the maximum permissible exposure. A loop antenna strung around the shack ceiling could create a strong electric or magnetic field around the operator or around the child sleeping (or playing) in the bedroom right above that ceiling.

If you will be using indoor antennas or antennas strung along the roof of your house, you should take steps to be sure that no one is subjected to more than the maximum permissible exposure limits of RF from your station. You can use measurements, computer modeling or data tables to perform such an analysis. The FCC Office of Engineering and Technology has published some information to help you with these evaluations in OET Bulletin 65 and Supplement B to that Bulletin (Additional Information for Amateur Radio Stations). You can find these documents on the World Wide Web at: **www.fcc.gov/oet/info/documents/bulletins**.

ARRL's *RF Exposure and You* includes information from those publications as well as extensive information and data tables to help you ensure that your station will meet the FCC requirements. — *Larry Wolfgang, WR1B, ARRL Senior Assistant Technical Editor*

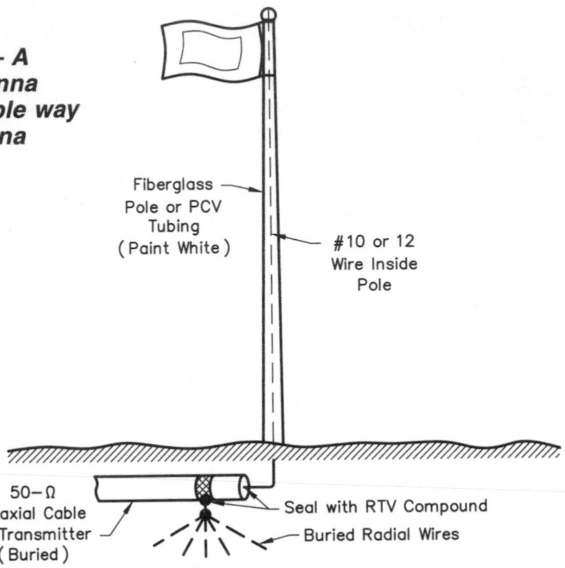

Figure 6-21 — A flagpole antenna can be a simple way around antenna restrictions.

fed wire antenna system if you are careful with the installation. One major drawback to an exterior invisible antenna using small wire is the effects of the wind on the small conductor. Invariably the wire will break and you will have to reaccomplish the installation. This option does allow the persistent QRPer to erect an outside antenna that will work very well, however. As with all end-fed antennas, don't forget the RF counterpoise.

If you live in a housing development that has restrictive antenna covenants, you might give some thought to erecting a decorative flagpole. There have been many articles in ham radio magazines attesting to the wisdom of using a flagpole as a vertical radiating element. The pole itself should be insulated from ground and you'll need to lay out a group of radials just under the sod. Fed via an antenna tuner with RG-213 that is buried in the ground, this "invisible" multiband antenna could be just the answer to getting on the air *and* showing the neighborhood your patriotic side at the same time. **Figure 6-21** shows one very simple installation of this type.

Another invisible antenna idea would be to use a "clothesline." **Figure 6-22** shows such an installation. Use Teflon or vi-

Figure 6-22—The clothesline antenna is more than it appears to be.

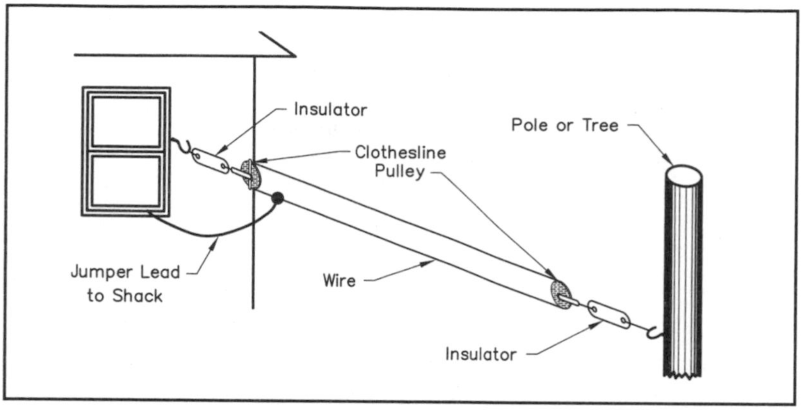

nyl covered #12 or 14 wire for the clothesline and you can probably even use it for *that* purpose. Use insulated pulleys.

One final thought about invisible antennas: many hobby shops that cater to people who work in stained glass or build doll houses have copper tape that has an adhesive backing. This stuff will work great for disguised antennas that are run along the eaves of the house. Place the tape down, use an SO-239 with pigtails and an alligator clip on each pigtail to clip to the tape, forming a dipole antenna. Once the tape is in place, a quick coat of matching paint and you cannot see this dipole at all! I would not try to run over 10 W to this type of antenna, due to the high impedance problems we have discussed previously.

The Oak Hills Research WM-2 QRP Wattmeter is a handy piece of test equipment. **(K7SZ Photo)**

TOOLS OF THE TRADE

Antenna measuring equipment can be as simple as a small Christmas tree light in series with the antenna (to show antenna current) or as complex as an antenna analyzer costing several hundred dollars. As with most things, what you need will lie somewhere in between these two extremes.

The most practical antenna measurement tool is the SWR bridge or forward/reflected RF power meter. There is very little difference in the two meter types. The SWR bridge is calibrated to read forward and reflected *voltage* while the RF power meter is calibrated in *watts*. Either type of meter will allow you to assess the condition of your feed line and antenna system, and tune it for the lowest *reflected* reading at resonance. There are several commercially made QRP meters available. I like to build things, however, and this much-needed accessory makes a very nice evening or weekend project. The Oak Hills Research WM-2 QRP Wattmeter (see Chapter 4) is accurate down into the milliwatt ranges.

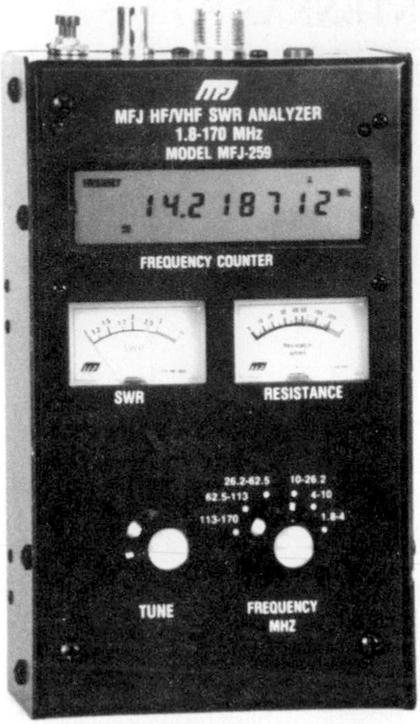

***The MFJ Model MFJ-259 antenna analyzer will help you measure the resonant frequency of your antenna system and make proper adjustments to the system. It will monitor SWR and even includes a frequency counter.* (Photo courtesy of MFJ)**

Antenna analyzers have become very popular over the last few years. I have used both the Autek RF-1 and the MFJ-259. I prefer the MFJ unit because it uses analog meters, which are much easier for me to read when trying to determining resonance of an antenna. Both units are quality products and perform a host of functions. The MFJ unit also functions as a VHF frequency counter, something the Autek unit does not. Neither of these two devices are cheap, so be prepared to shell out some money when you decide to buy one. The antenna analyzer makes life very easy for the antenna experimenter and should be considered if you plan on doing much antenna experimentation.

THE "LADDER-LINE" MULTI-BAND DIPOLE

Tom Hammond, NØSS, came up with this idea for a multi-band Field Day dipole antenna that was easy to store between outings ("Hints & Kinks," Sept 1998 *QST*, pg 78). I took Tom's idea and built several dual-band antennas for fixed station use. Of course, you could certainly stay with Tom's original idea and make several of these antennas for your Field Day and portable QRP operations.

The ladder-line I use is the 450 Ω variety that has a PVC coating. This feed line is available at most of the antenna hardware advertisers in *QST*. In this instance my ladder-line came from Jim Thompson, W4THU, at The Radio Works in Portsmouth, VA. pproximately 33 feet from one end and cut *only one side* of the ladder-line. Go to the opposite end and measure 33 feet down from that end and cut *the opposite side* of the ladder-line. Using a utility knife, remove the PVC insulation spacers between the two cuts. What you now have are two legs of a dual-band dipole, one element 65 feet in length and the other element 33 long, The ends that will attach to the feed point are stripped about one inch and soldered together.

For a center insulator I chose a $^{3}/_{4}$ inch PVC Schedule 40 pipe cap. Screw hooks are placed on opposite sides of the cap and a single screw hook is embedded directly in the top of the cap for center support as an inverted vee configuration. Additional holes are drilled just below where the screw hooks are attached to each

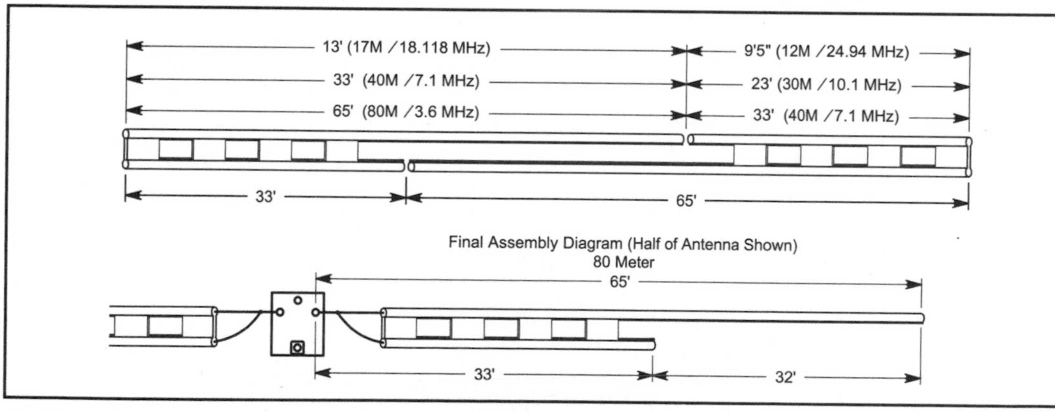

Cutting diagram for making an 80/40-meter dipole from 98 feet of transmitting ladder line. Dimensions are also shown for a 40/30-meter dipole and a 17/12-meter dipole.

MAKE ANTENNA CENTER INSULATORS FROM PVC PIPE CAPS

Photos A, B and C show three versions of a center insulator for a coax fed dipole. The center insulators are made from PVC slip caps, with the coax connectors and eyebolts mounted directly into the cap walls. These insulators are small, strong, lightweight and easy to build. They are built in $1^{1}/_{4}$-inch caps (outside diameter is about 2.0 inches). Close the open faces of the caps by gluing on a disc of sheet plastic. This sheet doesn't carry any load, it just seals and waterproofs the center insulator.

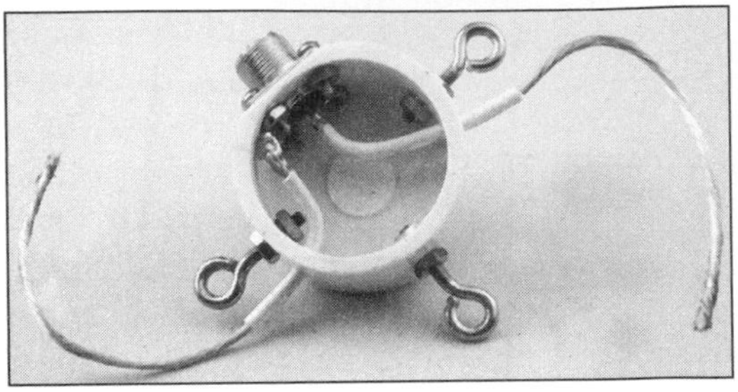

(A)

You can use several kinds of plastic sheet here. PVC sheet is available from hobby stores. Glue it to the PVC cap with PVC cement or vinyl glue. White polystyrene sheets are also available from hobby stores; glue them with coil dope from electronics distributors or with universal plastic cement (methylene chloride) from hobby stores. This cement will also attach Plexiglas sheet—available from hardware stores—to the PVC cap. This was done with the smaller center insulator in the photo, made from a $1^{1}/_{2}$ inch cap.

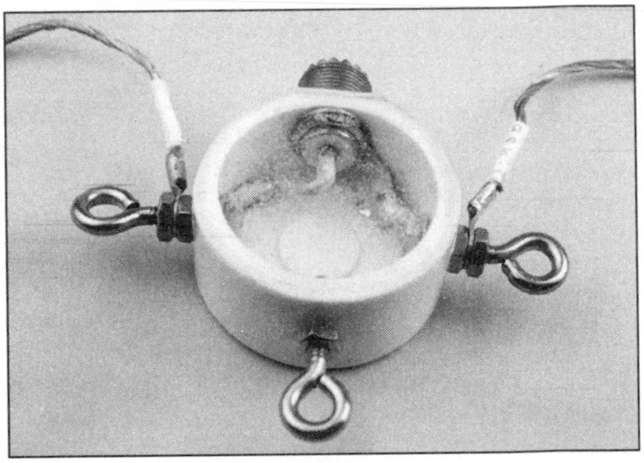

(B)

Use stainless steel or brass eyebolts and hardware for these insulators; plated steel will corrode. You can use PVC caps as manufactured, but the ones in the photos were cut down to an inside depth of about $1^{1}/_{16}$ inch, which comfortably holds the coax chassis-mount connector. Notice that an area on the side was filed flat to mount the coax connector.

Coax connects at the bottom of the insulators, and the antenna wires fasten to the two eyebolts at the sides. Stranded wire connects from the coax socket to the antenna wires. In two of the examples, the stranded wires exit the cap through holes beside the eyebolts. I sealed the

(C)

wire insulation to the cap where it exits the holes. In the third example, the wires connect the coax socket to solder lugs on the eyebolts *inside the cap.* Additional lugs and wires on the eyebolts *outside* the cap connect to the antenna wires. In all cases, wrap the stranded wires securely around the antenna wires, solder them and protect the joint against the weather. The eyebolt at the top is for an optional center support, as for an inverted **V**.—*Robert H. Johns, W3JIP, Box 662, Bryn Athyn, PA 19009*

W3JIP's idea suggests other applications for this technique. For multiwire dipoles, you could use a larger pipe cap (say, 4 inches, with pairs of eyebolts around the circumference for each band) or a section of pipe. See **Fig D** for some ideas.—*Robert Schetgen, KU7G, ARRL Hints and Kinks Editor*

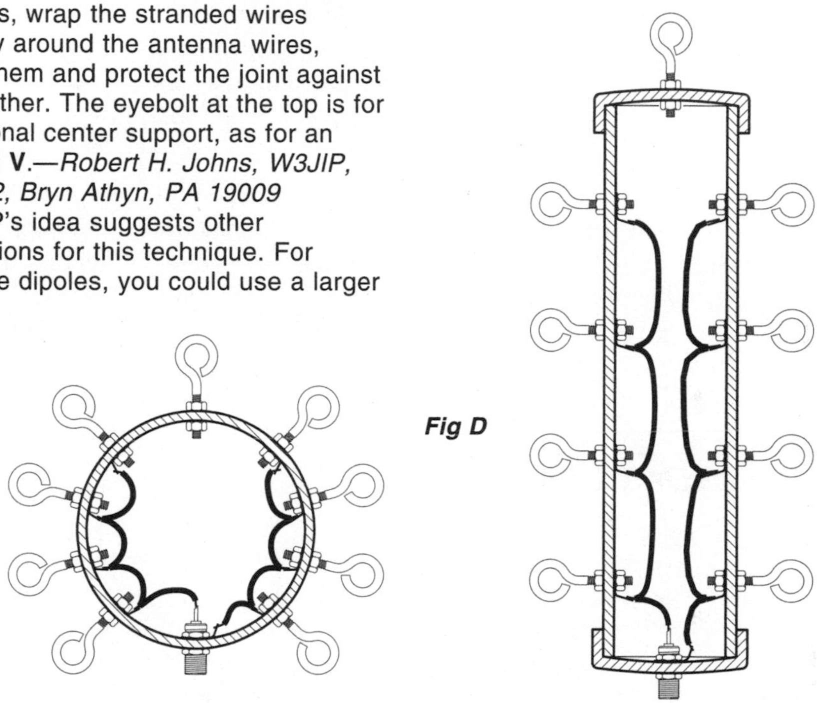

Fig D

side of the cap. An SO-239 coaxial connector is fitted with two pieces of wire, approximately 6 inches long, one to the center insulator and the other to the ground side of the connector. The wires from the connector are fed through the two holes in the cap and pulled taught until the connecter sits just above the open end of the pipe cap.

Using some 5-minute epoxy, I poured the epoxy into the pipe cap until it is just below the level of the open end of the cap. Now, I carefully wiggled the two wires protruding from the pipe cap until the coax connector nests into the top of the cap. At this point, the epoxy should be almost level with the edges of the pipe cap, with the rear of the coax connector firmly embedded into the epoxy. Allow the adhesive to fully harden and cure for 24 hours before hooking up the dipole elements. You now have a potted center insulator for your multi-band dipole.

To assemble the ladder-line dipole, insert the feed point ends of each element into their respective screw hooks. Clamp the end of each hook over so the elements cannot be disengaged from the center insulator. Now, take the end of each wire protruding from the cap and solder the ends to their respective elements. This results in an extremely robust antenna which will weather well in almost any climate.

Feed the center connector with 50 ohm coax and connect the other end to an antenna analyzer to check for resonant operation. Adjust the lengths of the dipole elements as needed to resonate the antenna on your favorite 80/40 meter QRP frequency.

Can this antenna be scaled for other frequencies? Most certainly. Just calculate the dimensions for your favorite bands and cut the ladder-line to the proper length.

THE APPALACHIAN TRAIL DIPOLE

Ed Breneiser, WA3WSJ, came up with a brilliant idea for a portable all-band antenna to take on hiking trips. Ed is among the rapidly growing number of QRPers who hike and operate from the Appalachian Trail all year long. Ed soon realized that height above ground didn't matter very much from the trail. Actually it was a moot point. The dense trees found along the AT often preclude having the antenna more than 15 feet off the ground.

The Appalachian Trail Dipole is essentially a 40 m dipole made using #26 AWG 19 strand Copperweld "stealth wire" and fed with twinlead, open wire or ladderline. The "stealth wire" is extremely rugged, with a non abrading PTFE outer coating that won't snag in trees and bushes.

The center insulator for the Appalachian-Trail Dipole is made in a similar manner to the aforementioned ladder-line multi-band dipole insulator, only on a smaller scale. A ¾ inch PVC Schedule 40 pipe cap is used with an SO-239 coaxial connector. You could use a BNC female connector here, but the miniature 300 ohm PVC ladder-line used as the feed line might pose a problem. See **Figure 6-23**.

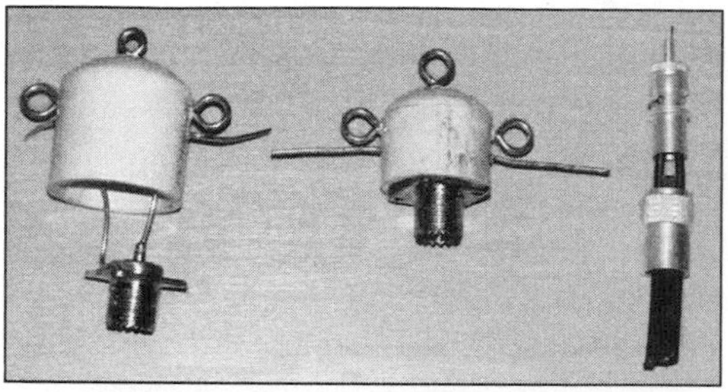

Figure 6-23 — The author's attempt at reconstructing WA3WSJ's multiband version of the AT Dipole. On the left is a 1-inch PVC pipe end cap, partially assembled. (The larger size makes the construction details a bit easier to see.) In the center there is a ¾-inch PVC end cap, fully epoxy potted and ready to use as a center insulator. On the right is a PL-259 connector with mini 300-ohm ladder line before soldering and potting with epoxy. **(K7SZ Photo)**

Instead of using screw hooks we use the "eyes" to secure the dipole ends and provide a center lift point for the inverted vee. Epoxy is poured into the pipe cap and the coax connecter is nested into the adhesive, after pulling the attached wires through holes in each side of the pipe cap just below where the screw eyes are attached. Once the epoxy has cured, it's a simple matter of attaching each 33-foot dipole element to the center insulator. The wire is looped through the eye and then knotted to provide strain relief. The wires from the SO-239 are then soldered to the pigtail of each element, providing continuity to the connector.

A PL-259 is used to terminate the miniature 300 Ω ladder-line to the center connector. The other end of this feed line is fed into a 4:1 balun and then into your favorite antenna tuner. I use an LDG Z-11 auto-tuner. **Figure 6-24** shows the LDG tuner and Yaesu FT-817

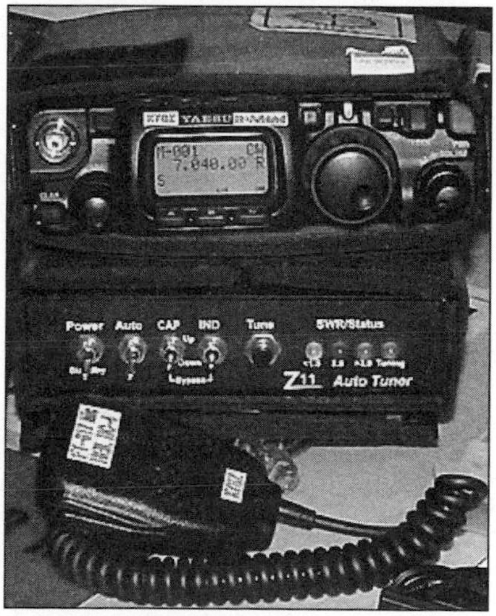

Figure 6-24 — The author's Yaesu FT-817 and LDG Z-11 Autotuner pack nicely into nylon carrying bags ready for portable operation. **(K7SZ Photo)**

ready to operate. Ed uses his K2 with internal auto-tuner, which easily handles tuning this 66 foot dipole on all bands *except* 17 meters. You can always use 50 Ω coax to feed this portable dipole and use the antenna as a standard 40 meter dipole.

The A-Trail Dipole is a very lightweight, extremely portable multi-band antenna system that will provide many hours of operation from the bush. Cost is minimal and the performance is amazing, even at low height!

THE PAC-12 PORTABLE VERTICAL ANTENNA

James Bennett, KA5VGS, has pioneered the development of an extremely successful portable antenna system called the PAC-12, which won the HFPack "Antenna Shoot Out" at Pacificon 2002. James' PAC-12 bested all comers including the Super Antennas MP-1, the usual assortment of single band whips and the Miracle Antenna. This "Shoot Out" has become a regular feature of Pacificon. The tests are conducted on a test range using a dipole reference antenna. All antennas under test are compared to this reference and are rated in descending order as to the amount of dB below the reference.

I was so intrigued by the PAC-12 that I decided to build one for myself and field test it on several outings over the summer. First of all, the majority of the components to construct this multi-band vertical antenna are available at most home improvement stores. The remainder is procured from RadioShack, so there is really no reason that a frugal QRPer should be without a good portable antenna. Total cost of this antenna is well under $25, lower if you have a good junque box.

I won't delve into construction details as a full account of how to construct the PAC-12 antenna is contained on the New Jersey QRP Club's "Website on CD ROM" available from the club for $10 on their website at: **www.njqrp.org**. After the unit was built, I used the MFJ-259 antenna analyzer to adjust the antenna coils on 40, 30, 20 and 17 m (these were the only bands I was really interested in for portable work). Each band coil assembly required only minor tweaking and final resonance was achieved by varying the length of the 6 foot whip at the top of the antenna.

James' design involves using two 12 inch sections of 1/4 inch aluminum rod, held together end-to-end with threaded couplings. The lower rod mates with the feed point insulator while the upper end

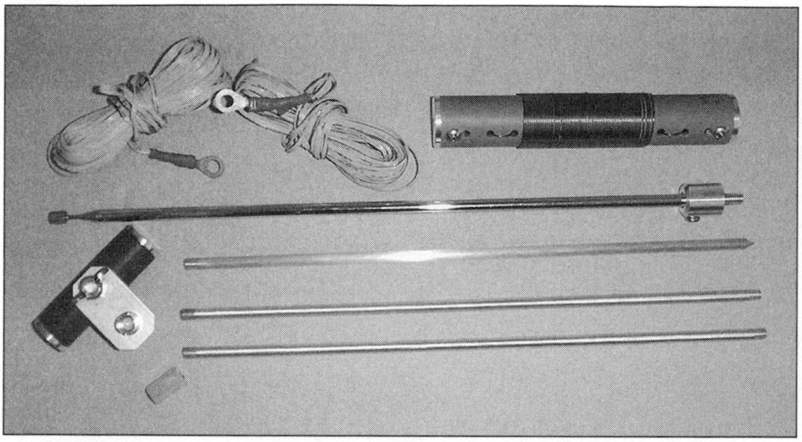

(A)

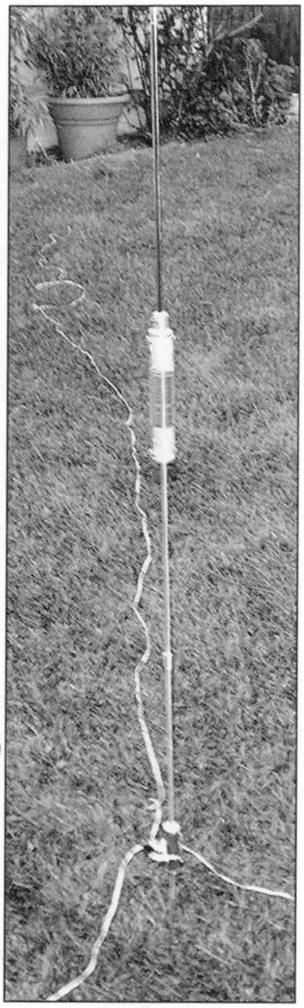

(B)

Figure 25 — Part A shows the author's version of the PAC-12 antenna ready to be packed for portable operation. Part B shows the antenna set up in the author's backyard, ready for some operating time. (K7SZ Photos)

mates with the band coil of your choice. The RadioShack six foot whip antenna sits atop the band coil making the overall height of this antenna around 9½ feet. Essentially this is a center loaded vertical. An additional 12 inch section mates with the bottom of the feed-point insulator and is either pushed into soft ground or held in a small tripod mount to raise the antenna about one foot or more above ground. Radials are attached to the ground end of the feed-point insulator. James recommends three to four radials, which can be made from virtually any small diameter flexible wire. Lay the radials out like spokes in a wheel around the antenna. The PAC-12 collapses into a compact bundle for easy transport. **Figure 6-25** A shows the pieces ready to pack in a small bag or box for transport. Part B shows the PAC-12 antenna set up in the backyard at K7SZ.

If your feed line run is less than 25 feet, RG-174 should work well. If you are paranoid about feed line losses, substitute RG-58U or RG-8X, both of which have less loss at higher frequencies.

My results with the PAC-12 were quite gratifying. Having experimented with a variety of portable commercial antennas over the last couple of years, it was nice to build an antenna for under \$25 that worked as good, if not better, than some the commercial versions I'd tested.

At this price, the frugal QRPer could easily build a second PAC-12 and, with the judicious use of phasing lines, produce a very potent two element vertical beam antenna. Hey, a portable phased two element 40 m array you can put in your back pack would make one heck of a Field Day antenna system!

The near-center-loaded design of this vertical results in improved efficiency when compared to base-loaded designs. I can readily see why the PAC-12 portable HF vertical antenna has found such favor with the QRP community. It is a solid performer that you can build for pennies on the dollar, in true QRP fashion. James now offers a commercial version of his PAC-12 in kit form. Check his Web site at **www.pacificantenna.com** or e-mail James at: **ka5dvs@pacificantenna.com**.

THE QUICK SILVER MINUTEMAN 20

"Where do you want to work from today" is the motto of the Quicksilver Radio *MinuteMan 20* portable HF vertical antenna. This portable backpack antenna is unique in that it is made from ¾ inch schedule 40 PVC pipe. It is lightweight (great for backpacking) and can be assembled in a few minutes. The MinuteMan 20 covers 20 through 10 meters and band changing is fairly quick, once the entire antenna is assembled. The antenna consists of six mast/leg sections, three base sections with "T" fittings attached, one upper assembly with collapsible 6 foot whip and loading coil attached, one lower assembly with coaxial cable connector, counterpoise connector and lower element wire attached, and two counterpoise cables each with two radials and terminals attached. This antenna is ready to assemble and use right out of the box. Quicksilver Radio included an excellent instruction set (owner's manual) which can get you on the air in minutes.

The MinuteMan 20 is a full ¼ wavelength antenna for 10, 12 and 15 meters. For operation on 17 and 20 meters, a small loading coil is placed in series with the vertical whip section. For 10, 12 and 15 meters, the lower element wire is connected to the *top* end of the coil and the whip is adjusted for best SWR. For

17 meters the lower element wire is clipped onto the center of the coil. To tune 20 meters the lower wire element is clipped to the bottom of the coil. Tuning is accomplished by adjusting the coil taps on 20 and 17 meters with the whip fully extended.

The MinuteMan 20 can be ground mounted or used on a picnic table or other elevated surface, thereby elevating the radials above ground to improve the antenna's efficiency. This is a fun antenna and a snap to assemble and use. It's bulkier than my and *PAC-12*, but band changing is a lot quicker. Go to **www.qsradio.com** and check out their full product line.

SUMMARY

In this chapter we have spent some time discussing the various types of antennas that are available to the QRPer. Emphasis was on simple, easily erected antennas that use wire as the main elements. I hope that this chapter has given you some basic ideas to try, and experiment with. Contrary to popular belief, you do not need a monobander at 120 feet to work DX. Simple wire antennas will work a lot of DX and get you into the DXCC.

As you progress in your QRP efforts, you will find yourself becoming involved with antenna experimentation. This is only natural, because the antenna is the one area that you can improve upon with very little cost. I would highly recommend a good antenna modeling software program. Personally, I like Roy Lewallen's *EZNEC* software. Several years ago Roy started producing a product called *ELNEC*. In 1997 he upgraded the analyzer program and renamed it *EZNEC*. Computer modeling using *EZNEC* allows you to build the antenna on the computer and sample how it should work *before* you erect it. It is much easier to change a few lines in the computer analysis than to cut-and-try when the antenna is up in the air. *EZNEC* runs on a 486DX66 or higher processor and is a great investment for the QRPer.

CHAPTER 7

HF Propagation for the QRPer

One interesting thing about writing the first edition of this book was to hear and read the feedback from the people who purchased it. According the readers, this chapter on propagation was one of the most informative and useful chapters. Therefore, it was an easy decision to include an updated propagation chapter in the new book. Understanding propagation is the key to success with QRP. Thanks to all who took the time to correspond with me regarding the book and especially this chapter.

My thanks also, to Tomas Hood, NW7US, for the in depth analysis on Solar Cycle 23. Tomas provided a wealth of background information on our current cycle (Cycle 23) and how it compared to previous cycles. His explanation and analysis of the huge solar flares in October and November of 2003 help put these record breaking solar flares into perspective and further help us to understand the propagation anomalies encountered during this time frame.

Since the first edition of this book, the peak of Solar Cycle 23 has come and gone. All QRPers were looking forward to Cycle 23, which officially started between May and June of 1996. Two separate peaks were observed in Cycle 23. The current cycle first peaked during April of 2000 with a smoothed sunspot number of 120.8. A second lower peak occurred in November of 2001 when the smoothed sunspot number was 115.6. Currently we are on the back side of Cycle 23 which is due to bottom out sometime between the end of 2006 and the beginning of 2007.

During this present cycle QRPers were hearing predictions

that Cycle 23 should reach the same magnitude as Cycle 21 with a monthly smoothed sunspot number reaching 160! Unfortunately, Cycle 23 compares more to past cycles 17 and 20, and to a lesser degree with Cycle 2, which all developed much the same way, with a smooth decline over a four year period before bottoming out and ending.

Although somewhat disappointing, QRPers experienced some great HF propagation yielding many openings to those rare areas of the world during Cycle23's peak period. QRP DXers were treated to a plethora of DXpeditions during Cycle 23, and many of us worked some killer DX during the top portions of this cycle.

While HF propagating was less than stellar, VHF propagation yielded some very pleasant surprises compared to past cycles. Memorable moments during Cycle 23 included intense periods of F-layer propagation, along with outstanding Radio Aurora and Sporadic-E openings.

Cycle 23 was also a history maker. This cycle provided us with a huge display of aurora thanks to some massive disruptions on the surface of our sun in the form of solar flares and accompanying coronal mass discharge (CME) events that far exceeded anything previously recorded! At the end of October 2003, we were treated to several large solar flares and CME events, along with the accompanying geomagnetic storms. While adversely affecting the ionosphere, these events provided some astounding visible aurora displays. Of course, auroral activity also equates to enhanced propagation on 6 meters. Needless to say, the 6 meter crowd was more than pleased that the solar flares had happened!

On November 4, 2003 **"The BIG One"** occurred. A solar flare on the sun's surface unleashed the largest X-ray event ever recorded. The sensors on solar monitoring stations were originally calibrated to include events up to a magnitude of 17.4. The blast of November 4th drove the X-ray sensors off scale for 13 minutes! Talk about spectacular! Using an interpretive method, scientists estimated that the magnitude of this event was between X25 and X28! Finally, the Space Environment Center (SEC) settled the argument and logged this event as an X28 flare, the largest flare ever encountered since records had been kept. Okay, big deal, the event was over 10 points higher than anything ever measured: on a *logarithmic* scale! Man, that's a *lot* of energy!

The ionosphere simply shut off. There was virtually no HF

propagation aside from ground wave and line of sight. It was like someone turned off "the big switch in the sky." I had heard rumors that the resulting aurora display was witnessed as far south as the Caribbean. What a light show! Of course, here in northeast Pennsylvania, it was overcast so I saw nothing but clouds and had to rely upon reports received via the VHF reflector and television! That's what it's like to be me!

While huge X-class flares of the magnitude that occurred on November 4, 2003 are not regular events, conventional thinking estimates that these probably occur every 20 to 40 years. We know one thing for sure: in the 30 year history of collecting solar flare data, **"The BIG One"** was nothing short of stunning!

Just because we are on the downside of cycle 23 doesn't mean that we need to put away the QRP rigs in favor of high power. While we won't enjoy super band openings on a regular basis for the next several years, enhanced HF propagation will still occur occasionally. Learning how to interpret the propagation information on WWV/WWVH at 18 minutes after the hour can give you a head start on working good DX. Likewise, reading and understanding the weekly propagation forecast bulletin from the ARRL will help prepare you in advance to face the foibles of HF propagation. Don't forget to check out Tomas Hood's propagation resource center on the Internet at **www.prop.hfradio.org**. Tomas also offers a propagation e-alert service on his website.

With out further ado, let's get to the meat of this chapter on propagation. What follows is probably the most important chapter in this book in order to achieve success as a QRP operator. There is a lot of information included in the next few pages, so plan on reading this chapter several times in order to maximize the usefulness of its contents.

To be successful in the low power arena, you must have a thorough understanding of the ionosphere and how propagation of HF signals takes place. The technical information in this chapter can seem pretty intense. To be sure, we will only scratch the surface of HF propagation in this chapter, but I will refer you to other reading that will provide more in-depth coverage of the topic.

In this chapter you will learn many new terms, theories and abstract concepts that will, at first, seem to be overwhelming. Seasoned amateurs will find this chapter a good brush-up on propagation theory. To the neophyte: do not become discouraged. Take your time and reread this chapter several times. Refer to *The New Shortwave*

Propagation Handbook by George Jacobs, W3ASK; Theodore J. Cohen, N4XX; and Robert B. Rose, K6GKU, for an in-depth analysis of the propagation phenomenon.

QRPers need to understand all aspects of the ionosphere in order to communicate effectively over great distances. It is via the ionosphere that high frequency (HF) communications is made possible. In this chapter we are going to discuss some of the ways that HF waves propagate around the Earth, how best to calculate which bands are open to various areas of the world, what the WWV propagation forecasts tell us and how to use this information effectively.

Radio waves in the MF and lower HF range (ie, 160 and 80 m) tend to travel well along the surface of the Earth. This is referred to as *ground-wave propagation*. All radio waves have a ground wave. MF signals (especially the AM Standard Broadcast Band and 160 m) tend to have a much longer ground-wave distance since they encounter less absorption by the Earth's surface. As the frequency is increased above about 7 MHz (40 m), the ground wave exists for only a few miles. Any communications out past the ground wave must rely on the *direct wave* (line of sight between the transmitter and the receiver) or *sky wave* via *ionospheric propagation* (refraction of an airborne HF radio wave by the upper levels of the ionosphere).

RF energy between 2 and 50 MHz may be refracted, or bent, by the ionosphere depending upon the time of day, the angle at which the RF wavefront enters the ionized layer (called the *angle of incidence*) and the degree of ionization present in the refracting layer. This same wavefront could also penetrate the ionized layer without being refracted if the angle is not correct or the ionization level is low. Additionally, if the ionization level is extremely high, the HF radio wave will be almost totally absorbed. If you are beginning to see that HF propagation is a high tech "crap shoot" you are starting to get the point of this chapter.

The ionosphere, as we refer to it in communications, extends from about 30 miles to approximately 400 miles above the surface of the Earth. This 370 miles of atmospheric area is separated into several zones or layers, which are constantly changing, depending on the time of day, season, year and ultraviolet emissions from our Sun. **Figure 7-1** shows how these zones change with season and time of day. Since the air closest to the surface of the Earth (tropospheric layer) is quite dense, the Sun's ultraviolet

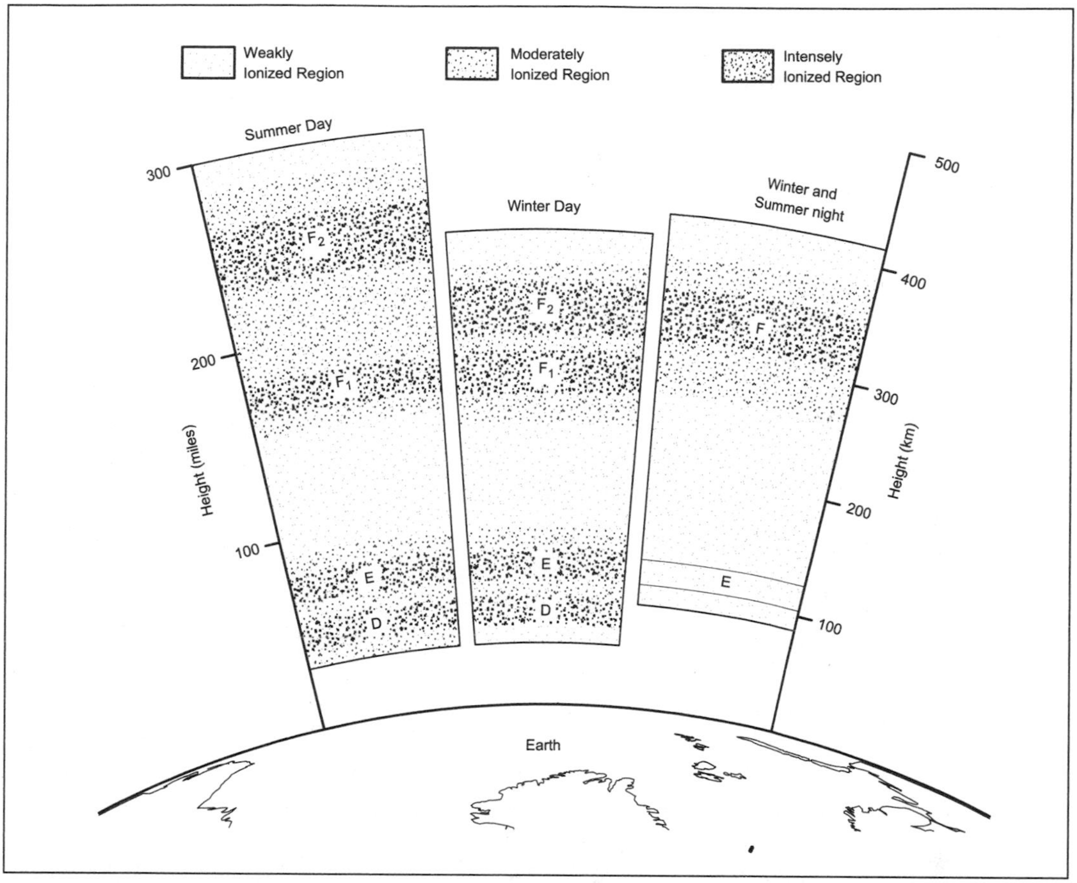

Figure 7-1 — This diagram illustrates the daily and seasonal variations in the ionospheric regions. Typical characteristics are shown. Note that there is some residual ionization in the E layer during nighttime hours. [Adapted from: George Jacobs, et al,* The New Shortwave Propagation Handbook *by permission (CQ Communications, Inc, 1995), p 1-8.]

radiation has little effect on the formation of ions within the troposphere. When we enter the lower layer of the ionosphere, however, (starting about 30 miles above the Earth) the density of the molecules is much lower, and the Sun's ultraviolet emissions can start ionizing the gas molecules. As we proceed upward from 30 miles to about 400 miles, the Sun's ionizing effects become much more pronounced.

In actuality, the process happens in reverse order, starting with the outer layers of the ionosphere and descending downward toward the troposphere. As ultraviolet rays emitted from the surface of our Sun encounter the very thin upper levels of the ionosphere, electrons in the outer shells of the gas molecules (mainly hydrogen and helium) are torn loose from their orbits, creating positively charged hydrogen and helium ions along with negatively charged free electrons. This process is called *ionization*. This ionized area is quite effective in bending, or *refracting*, RF radiation. During nighttime, this process is reversed. Called *recombination*, it occurs when the free electrons are recaptured, forming stable gas atoms.

As the ultraviolet radiation from our Sun encounters the first layers of the upper ionosphere, gas ions are created. As the UV radiation continues downward through the ionosphere, it encounters more dense layers of gases and the ionization effect is increased. The deeper the penetration of the UV radiation, the more highly ionized the ionospheric gases become. Eventually, the UV penetrates so far into the dense layers of the ionosphere that it starts losing energy and can no longer sustain the ionization process. The overall result is an intense layer (or layers) of ionized gases with areas of diminishing ionization above and below them. This area of ionization can bend or refract HF radio waves very effectively, enabling long distance communications to take place.

The gases in the ionosphere respond to different frequencies across the ultraviolet spectrum. This results in layers of dense ionization being formed at several definite heights above the Earth. The whole process forms the layers of the ionosphere. The amount of ionization in a particular region depends on the intensity of the UV radiation striking that area.

IONIZED LAYERS

The Sun's UV radiation produces several distinctly different layers: The *D layer* (approximately 30 to 55 miles above the Earth), the *E layer* (approximately 55 to 75 miles above the Earth), and the *F layer* (approximately 90 to 300 miles above the Earth). The F layer actually splits into two distinct regions during daylight: the *F1* and *F2 layers*. Each of these layers has unique characteristics that dramatically affect their ability to bend (refract) HF radio waves depending on the time of day, season of the year and current sunspot activity. Figure 7-1 is a diagram of the Earth and the four ionospheric layers that can refract radio waves.

How does an HF radio signal travel via ionospheric propaga-

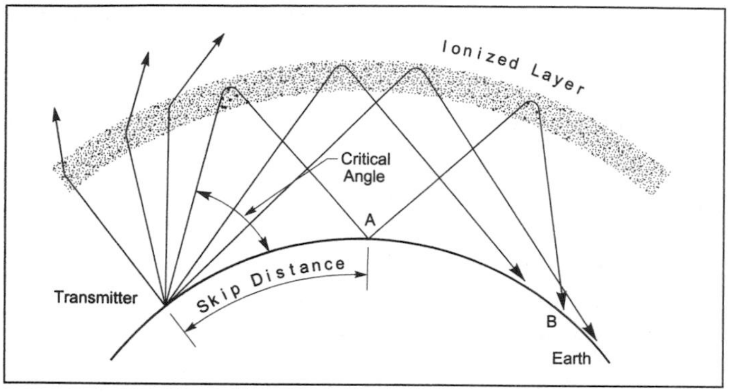

Figure 7-2 — The behavior of waves encountering the ionosphere is shown here. Rays entering the ionized region at angles above the critical angle are not bent enough to be returned to Earth, and are lost to space. Waves entering at angles below the critical angle reach the Earth at increasingly greater distances as the launch angle approaches the horizontal. The maximum distance that may normally be covered in a single hop is about 2500 miles. Greater distances are covered with multiple hops.

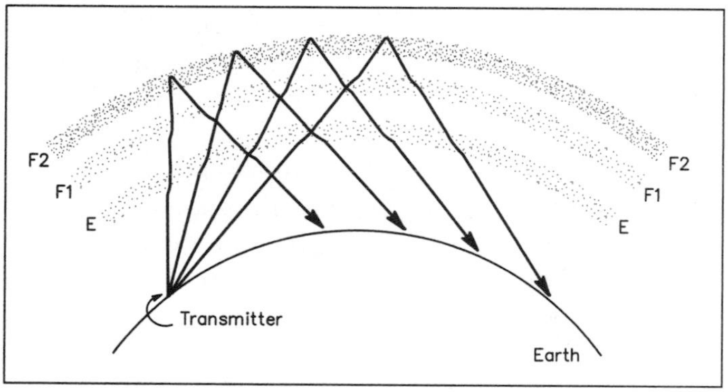

Figure 7-3 — Here we see typical daytime propagation of high frequency (14 to 28 MHz) signals. The waves are partially bent going through the E and F1 layers, but not enough to be returned to Earth. The F2 layer bends the signals enough to return them to Earth.

tion? The answer is quite simple. As the HF radio wave leaves the antenna, the ionized upper levels of the atmosphere (remember, it is less dense than air closer to the Earth) allow the radio wave to travel faster than the nonionized lower air. This results in the top part of the RF wavefront moving into the ionosphere ahead of the lower part of the wavefront. Eventually, the wavefront may turn (if conditions are just right) and refract or be bent back downward toward the surface of the Earth. This bending or refracting is the critical part that the ionosphere plays in HF propagation. As the RF wavefront enters the ionosphere and is refracted back down to Earth, the distance between the two points is referred to as *skip distance* or *skip zone*. This distance is directly related to the *angle of refraction* caused by the ionized gas layer. This angle will vary between daylight and darkness, time of day, season and frequencies used. **Figure 7-2** shows a simple example of radio waves being refracted and returned to Earth from a single ionospheric layer. **Figure 7-3** illustrates the bending that occurs as the signal passes through the lower layers of the ionosphere and the return to Earth from the F2 layer.

It is important to remember that only a very small portion of the transmitted RF wave is actually returned to Earth at any one point. Add to this the fact that really long-haul HF communications requires several skips or return trips up and down between the ionosphere and the surface of the Earth (this is called *multihop propagation*), and you get the feel for the obvious: the returned RF wave on a multihop path is extremely small compared to the original radiated signal.

Below the D layer, but above the actual breathable atmospheric layer, exists a region called the *troposphere*. Here, HF radiation is largely unaffected by the dense air molecules. HF signals are only slightly absorbed by the troposphere on their way through to the ionosphere. VHF and UHF radio waves, upon entering the troposphere, can become enhanced and travel many hundreds of miles by encountering a phenomenon called *tropospheric ducting*. Tropo ducting causes VHF/UHF signals to propagate hundreds or thousands of miles. This phenomenon accounts for television viewers on the East Coast occasionally being able to watch stations from the Midwest.

The US military has, for many years, used the troposphere to provide a means for reliable tactical and long haul point-to-point communications via their high powered tropo-scatter systems. **Figure 7-4** shows how a tropospheric scattering region can extend

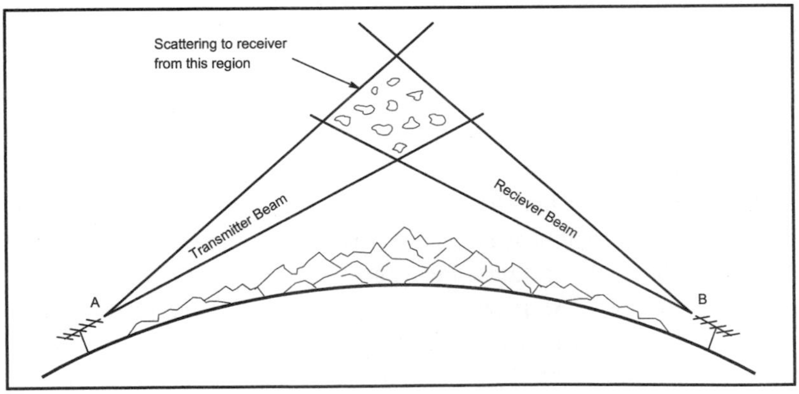

Figure 7-4 — Tropospheric scattering occurs from random "blobs" of air in the area where the transmitting and receiving beams intersect.

the range of VHF and UHF stations.

Although it seems that this whole idea of ionospheric propagation is rather hit and miss, many of the ionosphere's characteristics are predictable. First we will study the different layers and how and when they form.

THE D LAYER

The D layer is the lowest layer of the ionosphere. It can become readily ionized by the Sun's ultraviolet emissions. As stated earlier, the D layer exists between about 30 and 55 miles above the surface of the Earth. It forms quickly after local sunrise and reaches its peak ionization around local noon. Shortly after sunset, the ions in the D layer recombine to form gas atoms, and the layer disappears. HF signals are not bent enough by the D layer to return them to Earth, so the signals normally pass completely through this layer. The D layer does manage to absorb some energy from any HF signal that passes through it, however. During the day, when the layer has the highest ionization, signals on the lower HF bands (160, 80 and 40 m) are almost totally absorbed, so those bands are not generally useful for long distance communications during the day. Ionospheric storms can increase the D layer absorption to the point that no signals pass through this layer on to the upper levels of the ionosphere. The HF bands can seem to go dead, with no long distance propagation during such times.

THE E LAYER

The E layer exists from about 55 to 75 miles above the surface of the Earth, mainly during the daylight hours. This layer is the first effective area that HF radio waves can encounter and be refracted back toward Earth. The actual height of the E layer remains relatively constant during daylight hours but may vary during seasonal changes. The level of ionization is very high compared to the D layer and follows the movement of the Sun across the sky, with the maximum ionization occurring around local noon. Shortly after local sunset the E layer disappears almost completely. The area between the D and E layers is an area of reduced ionization. There is no real separation between the two layers, however.

THE F LAYER

The F layer is unique since it splits into two separate ionized layers during daylight. The F1 layer exists from about 90 to 150 miles above the Earth's surface. The F2 layer extends from 200 to 300 miles up. Notice that there are distinct areas between the two F layers and the E layer. The F layers are the two most important regions for the HF communicator. The F1 layer exists above and is much more highly ionized than the E layer. This layer exhibits much the same behavior as the E layer, forming after local sunrise, attaining maximum ionization around local noon and recombining to almost nothing during hours of darkness. At night the F1 layer recombines leaving only the F2 layer, which descends to between 150 and 250 miles above the surface of the Earth. The F2 layer is present all year long, day and night. It is the only ionization layer that is always in existence, irrespective of time of day. The F2 region has the highest of all ionization levels and is in a constant state of flux. Since the recombination of ions takes place slowly in the F2 layer, this region is always ionized to some extent. If it was not, nighttime skywave propagation of HF radio signals would be impossible. The F2 layer moves about constantly, changing height and ionization intensity between daylight and darkness.

SPORADIC-E PHENOMENON

No discussion of HF propagation would be complete without mention of the very unpredictable *sporadic-E* propagation. For reasons not fully understood, highly mobile, very thin clouds of ionization occur, apparently at random, at about 60 miles in height. Since this is in the E region of ionization, this phenomenon is referred to as sporadic-E. These highly ionized, fast moving clouds of gas mainly occur during the summertime over a specific geographic area. In addition, they can also occur less frequently at nighttime during the winter months. These wintertime displays of sporadic-E propagation are often associated with displays of northern lights or aurora borealis. The overall effect of a sporadic-E cloud is a sudden increase of useable frequencies, often going to 28 MHz and above. During a sporadic-E event you will notice that propagation on 15 and sometimes 10 m becomes highly enhanced in a given geographic area, allowing communications over paths not normally open at that particular time. This

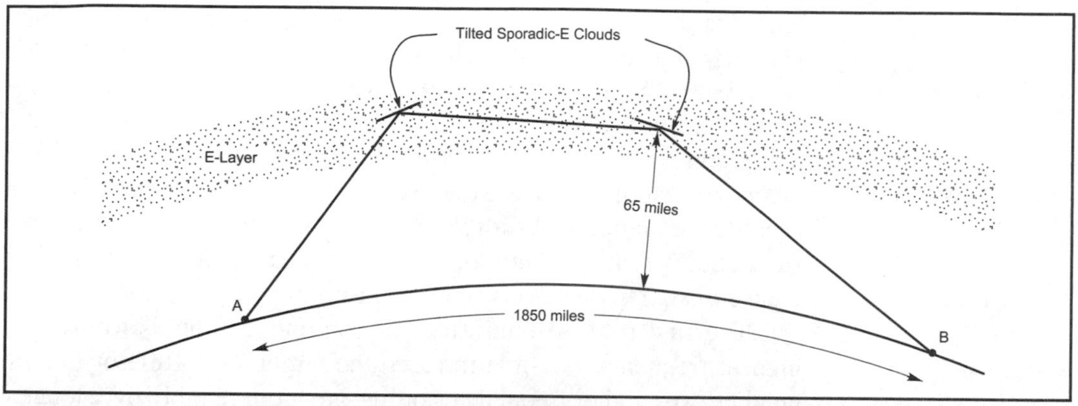

Figure 7-5 — This drawing shows a proposed explanation for observed 1400 to 2000-mile sporadic-E contacts that behave as if propagated via single-reflection paths. Distances longer than normal single-reflection paths might be possible by means of reflections between tilted E clouds. The MUF of the sporadic-E clouds along an Earth-cloud-cloud-Earth path need not be as great as that for the cloud in a single-reflection Earth-cloud-Earth path because the reflection angles required to bring signals back to Earth are less for the Earth-cloud-cloud-Earth model. [Based on Adolf K. Paul, "Limitations and Possible Improvements of Ionospheric Models for Radio Propagation: Effects of Sporadic E Layers," **Radio Science,** ***21 (May-June 1986), pp 304-308.]***

is an effective way to contact close-in stations on 10 m that would otherwise remain unworked. See **Figure 7-5**.

PROPAGATION VARIABLES

As you can see, the four layers of ionization are responsible for HF radio wave propagation over extremely great distances. These layers are dramatically affected by several predictable variables, namely: diurnal-nocturnal (day-night), seasonal, geographical and cyclical variations of the Sun. We will look at each of these propagation variables in detail.

DIURNAL-NOCTURNAL TRANSITIONS

The day-night transitions are caused by the rotation of the Earth with respect to our Sun. During the daylight hours, intense ultraviolet radiation from the Sun causes the various layers of the ionosphere to become ionized and support HF radio communications. At night almost no UV from the Sun finds its way to our atmosphere and the ions in the various propagation layers recombine, leaving only the F2 layer to support propagation. The diurnal variations are directly related to the position of the Sun in the sky. As the Sun approaches zenith (local noon), ionization increases proportionally and the frequencies used to communicate along a given path also increase. The intense ionization in the F2 layer from sunrise through sunset is reflected by an increase in useable frequencies. As ionization increases in the F2 region, the frequencies that will support long-distance communications increase also. This is a very important point to remember.

The first rule of thumb for HF communications is to use the highest frequency possible to reach the target area. Rule of thumb number two: higher frequencies propagate more reliably over longer paths. A station on the East Coast may be able to only work stations in neighboring states (about 300 miles) on 40 m during the midmorning. During this same period, however, the East Coast station can be reasonably assured of communicating with stations on the West Coast (over 2500 miles distant) by switching to frequencies in the 10 or 15 m bands.

Each refractive layer has its own *critical frequency.* This is the highest frequency that will be refracted by that particular layer. The critical frequency is constantly changing due to diurnal-nocturnal, seasonal and geographical variations, which will be discussed in the following paragraphs. This critical frequency is the prime indicator of the physical characteristics of the various ionized layers in the ionosphere.

SEASONAL VARIATIONS

Seasonal variations occur because of the ever-changing position of the Earth in relationship with our Sun. During the winter months, the Northern Hemisphere of the Earth is tilted away from the Sun resulting in the Sun appearing to be lower in the noontime sky. In the summer months exactly the reverse occurs and the Sun appears higher in the noontime sky. Atmospheric ionization is much higher in the summer as opposed to the winter months because the Sun is higher in the sky and there are more daylight hours. Seasonal variations in the E layer are very predictable, closely following the elevation of the Sun in the sky.

The F layer is the prime supporter of HF long distance communications, and it exhibits some rather unusual characteristics during seasonal variations. During the spring, summer and fall, the F1 layer closely tracks the performance of the E layer, with intense ionization occurring at local noon. During winter, however, the F1 layer merges with the F2 layer and HF long haul propagation is entirely dependent on the F2 region. The performance of the F2 layer during wintertime is somewhat complicated. During daylight hours the level of ionization is quite high and therefore, the critical frequencies are also high. However, during the extended hours of darkness the F2 layer has longer to recombine and ionization levels drop radically. This, in turn, causes nighttime critical frequencies to become lower.

Conversely, during the summer hours of daylight, the intense heat from the Sun's rays causes an expansion of the F2 layer gases. This results in much lower ionization levels (due to much less dense accumulation of gas molecules) and a corresponding decrease in critical frequencies. Since summer nights are shorter than winter nights, the F2 layer has less time to recombine and consequently the nighttime critical frequencies are considerably higher than in winter. In actuality, the difference between daytime and nighttime critical frequencies during the summer are small compared to the wintertime.

So what does all this mean? Briefly, during the daylight hours the E layer will support short-skip HF propagation. At night, the F2 layer is the sole refractive region. During the wintertime the F2 layer will provide long skip or multihop propagation paths to various areas of the world during the daylight hours. On winter nights, the F2 layer will continue to support HF communications, although the frequencies will be slightly lower than during the daylight hours. In the summer, the F2 layer will provide communications paths during daylight on frequencies that are lower than their wintertime counterparts. During the hours of summer darkness, the F2 layer will support HF communications on much higher frequencies than during the winter.

GEOGRAPHICAL VARIATIONS

Geographical variations that affect the refracting layers produce an interesting phenomenon. Since the Sun is directly overhead at the equator, ionization along the equatorial regions is considerably higher than other geographical areas, especially the high northern and southern latitudes. E layer and F1 layer critical frequencies are directly proportional to the Sun's elevation in the sky. Hence, during daylight in the equatorial regions, ionization is much more intense in these regions and therefore, the critical frequencies are proportionally higher. Above and below the equatorial belt the level of ionization tends to drop off proportionally with the subsequent drop in critical frequencies.

Likewise, the F2 layer critical frequency also tends to follow the general pattern of more intense ionization in the equatorial regions, with a decrease in ionization above and below the equatorial belt. In addition to the latitudinal variations, the F2 layer also tends to vary in intensity along certain lines of longitude. The generally accepted theory for this phenomenon has to do with the variations in the Earth's magnetic field. The F2 layer critical frequencies are generally higher in the regions of Asia and Australia than in Africa, Europe or the Western Hemisphere.

CYCLICAL VARIATIONS OF THE SUN

Cyclical variations of the surface of our Sun also drastically affect HF propagation. Every 11 years (roughly) there is a start of a new sunspot cycle on the Sun's surface. This 11-year sunspot cycle has been tracked and activity recorded for many years. The number of sunspots (large explosive disruptions on the Sun's surface, which produce huge amounts of ultraviolet radiation and X-ray emissions) enhance the ionization of our upper ionosphere. The sunspot activity changes dramatically during the 11-year cycle. When maximum sunspot activity is observed we say that we are encountering the peak of the solar cycle. Conversely, when the minimum activity occurs we are at the bottom of the solar cycle. In between, you have a very steep rise from the bottom of the cycle to the peak (which usually takes about four years) and a much slower decline from the peak back down to the bottom of the cycle (about six to seven years). The actual peak will vary from six months to one year. We began to see the increase in sunspot activity that signaled the start of Solar Cycle 23 in about April 1997. Over the three years following the start of the cycle, we can expect to see a dramatic increase in propagation, especially on 15 and 10 m, two of our best QRP bands.

The 11-year solar cycle has the most profound effect on the F2 layer. Critical frequencies at local noon will be about twice as high during the solar cycle peak as compared to the bottom of the cycle! Nighttime critical frequencies tend to track the daytime highs with the critical frequencies at local midnight about twice

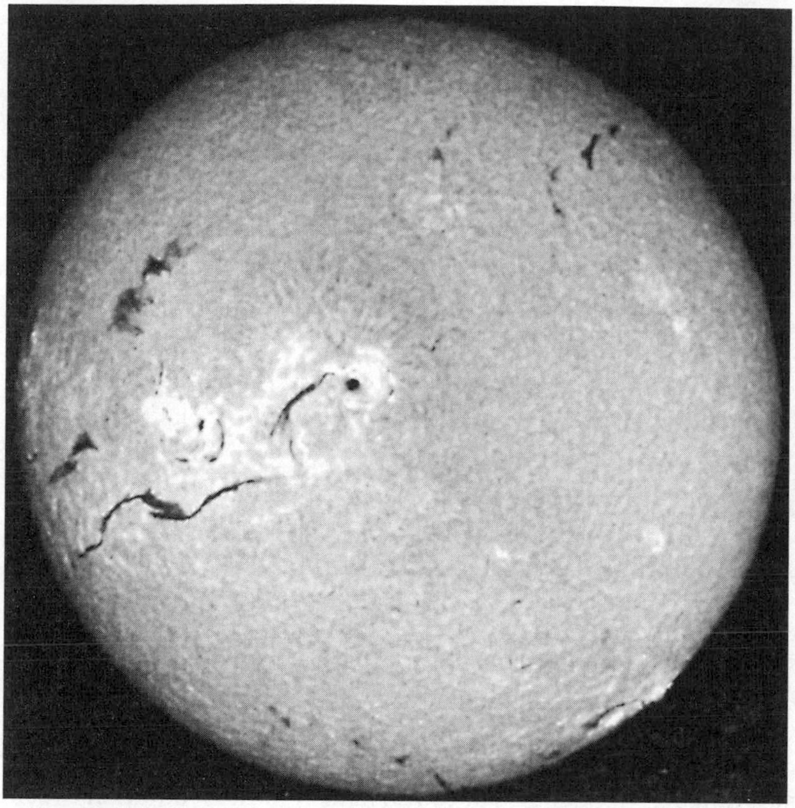

Much more than sunspots can be seen when the sun is viewed through selective optical filters. This photo was taken through a hydrogen-alpha filter that passes a narrow light segment at 6562 angstroms. The bright patches are active areas around and often between sunspots. Dark, irregular lines are filaments of activity having no central core. Faint magnetic field lines are visible around a large sun-spot group near the disc center. **(Photo courtesy of Sacramento Peak Observatory, Sunspot, New Mexico.)**

as high on the cycle peak as compared to the cycle minimum.

The E and F1 layers also show marked increases in ionization and corresponding critical frequencies during the peak of the solar cycle. Variations between the peak and the bottom of the cycle are not as pronounced as with the F2 layer.

The previous information has been included to show the interrelationships between our ionosphere, seasons, day/night transitions, cyclical variations of the sun and HF propagation.

MUF, LUF AND FOT

We need to define several terms before we move on. The first term is *maximum useable frequency*, or MUF. The critical frequency discussed earlier is obtained from firing an RF signal vertically into the ionosphere. Each layer will return some frequency, and when the highest frequency for each layer is recorded, that is

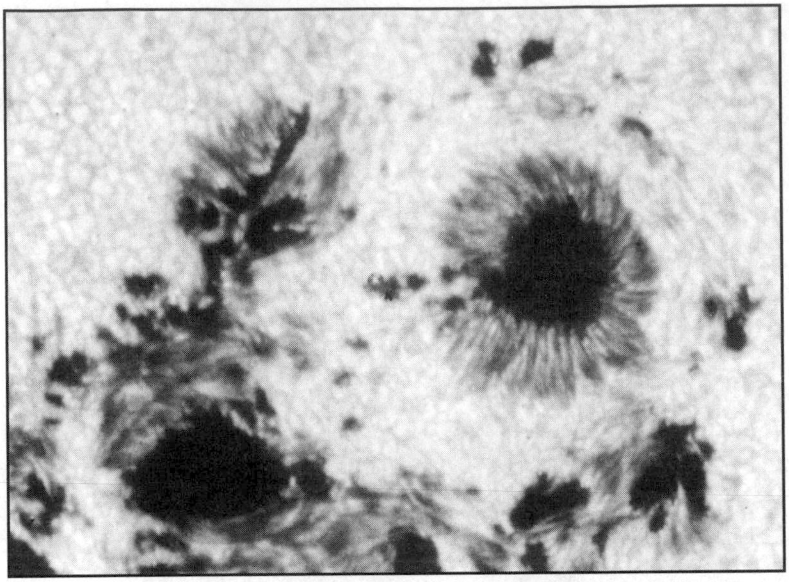

***Large sunspots embedded in the solar surface are shown here. Photographed with unprecedented sharpness, this shot was taken from an unmanned research balloon at an altitude of 80,000 feet. Note the granular composition of the Sun's surface.* (Official National Science Foundation photo.)**

known as the *critical frequency* for that particular layer. The MUF is obtained from a trigonometric equation that uses the critical frequency for a particular layer and the distance between the transmitter and receiver sites. Once computed, the MUF indicates the *maximum* frequency that can sustain communications between the two locations for a given time of day, season and layer.

In order to communicate from point A to point B, the frequency chosen must be equal to or less than the MUF. Using the MUF does not guarantee success. Remember, this is the *maximum* frequency that will propagate between the two points for that particular layer. One interesting note: the radiated power does not figure into the MUF equation. The RF wave will be refracted by a given layer based solely on the intensity of the ionization and the frequency of the radio wave. Hence, QRP radio signals have just as good a chance of propagating around the Earth as higher power signals! This is a real confidence builder!

The next term that needs to be defined is the *frequency of optimum traffic*, or FOT. The FOT is the most favorable frequency that can be used between two locations. The FOT will provide the most reliable path and be less subject to fading and other propagation anomalies. The FOT is calculated at 85% of the MUF (MUF × 0.85). The FOT, by definition, will support communications between two locations on 90% of the days of the month provided it is above the LUF.

The *lowest useable frequency*, or LUF, is the lowest frequency that can be used over a given path at a given time to support communications. By definition, LUF is the frequency at which the received signal strength is equal to the minimum signal strength required for satisfactory reception. Minimum signal strength depends upon transmitter power, gain of the transmitting antenna, path length, ionospheric absorption and gain of the receiving antennas. Lots of variables here, but notice that this is the first time *transmitter power* is brought into the equation. Other factors that affect the received signal are noise at the receiving site, and *type of modulation used*. Here is where CW tends to outshine SSB and FM as the favored mode of transmission. It is much easier to distinguish dots and dashes against background noise as opposed to speech using SSB or FM. In fact, CW requires a signal-to-noise ratio of only 3:1 compared to an SSB signal-to-noise ratio of 7:1. Simply put, a copiable CW signal will result if the received energy is only three times higher than the noise at the receiving station. A readable SSB signal needs to be seven times higher than the background noise to be copiable.

BAND-BY-BAND TOUR

Now let's take a look at the various HF bands and what the QRPer can expect as far as propagation is concerned. Remember, the following information is presented as a guideline only. Actual propagation conditions will vary and, in some cases, vary quite drastically from predictions. Remember also that the propagation follows the direction of the Sun from east to west.

160 METERS

The 160 m band is a nighttime band only. During daylight hours, signals are absorbed by the D layer, and the band is dead. (You may be able to make contacts out to 100 miles or so using ground-wave propagation.) Shortly after local sunset, however, the signals start to rise out of the background noise and propagation out to about 1000 to 2300 miles is possible. Efficient antennas are the name of the game on 160. Signals peak at about local midnight and then start to decrease. By dawn the band is about closed. A pre-sunrise/sunset period does exist for grayline DX propagation. To be effective, both sites must be either in transition from dawn to dusk or vice versa. Winter is the best time for working 160, as the band noise is relatively low. In the summertime 160 provides some contacts but conditions are limited due to atmospheric noise and thunderstorm activity.

80 METERS

Eighty meters is also a nighttime band. This band can provide some action during hours of daylight as long as absorption is not intense. Distances will be limited to perhaps 250 miles or so during the day. Like 160, signals on 80 m come up out of the noise about local sunset and continue to rise until local midnight. The signal levels are fairly constant from midnight to local dawn, when they start dropping off into the noise. This is a good QRP band, with the East Coast ops having a good shot into Europe and the West Coast QRPers having a shot into the South Pacific and Japan during the nighttime hours.

In order to work DX, efficient antennas are a must. Local QRP contacts (within 200 to 300 miles) can be made using only a simple dipole. Occasionally really long haul DX is possible using compromise antennas. This is the exception rather than the norm, however. My experiences on 80 m tend to make me believe that unless you have a full size dipole at least 60 feet high, your chances of doing anything tremendous on this low HF band are greatly reduced. The 80 m band becomes a valuable DX band during the bottom of the sunspot cycle when all critical frequencies are lower.

40 METERS

The 40 m band is by far the most popular QRP band. It tends to follow the ionization levels very closely. The performance during daylight hours will yield distances of about 500 miles. Nighttime is where things *really* start to happen on 40, though. Signals on 40 m rise quickly after dusk. Then some strange things start to happen. Recombination of the F layer results in some very unsettled conditions until about 2200 to 2300 local time. Once the F layer settles down, transcontinental communications are possible on 40! The lowly dipole, suspended at 25 to 40 feet, is more than capable of working local and DX contacts on this band. Obviously the higher the antenna the more efficient it becomes, therefore, an optimum antenna height of 45 to 50 feet will produce outstanding results on 40 m using QRP power levels. During QRP contests 40 m is *the* band. This band consistently produces the bread-and-butter contacts for my contesting efforts. If a single-band contest entry was contemplated, 40 m would be my choice. Even mediocre antennas perform well on this band. Loop antennas are very effective on 40 and yield much quieter background noise.

30 METERS

The 30 m band is great for QRP operation. Low population coupled with conditions that mirror both 40 and 20 m blend to make this band a QRPer's dream. The 30 m band is open both during daylight and at nighttime. Summer noise levels are lower than on 40 but absorption is greater than on 20 m. Diurnal/nocturnal shifts are not as pronounced as on the higher-frequency HF bands. Be ready for the rapid shifting of propagation conditions that are prominent on 40 m after local sunset. Power on this band is also limited, making the QRPer much more competitive. Dipole antennas are quite common and will yield many QRP contacts, even under depressed MUF conditions.

20 METERS

The 20 m band is the lowest frequency of the DX bands. It is open most of the time to various areas of the world. Truly long haul QRP contacts can be made on this band, depending on seasonal, diurnal/nocturnal and cyclical variations. The 20 m band closely follows the ionization levels in the upper ionosphere. Critical frequencies increase steadily from dawn to about local noon. Here they stabilize with a slow return during recombination of the F layer ions after local sunset. From spring through fall 20 m will remain open almost 24 hours per day during much of the sunspot cycle. During winter, propagation will diminish with early nightfall. During the bottom of the sunspot cycle, 20 is often the only band open to many DX locations. This makes it *the* DX band during periods of low sunspot activity. Using a 20 m dipole 25 to 30 feet high will yield DX contacts on a regular basis. Obviously, 20 m is a prime candidate for a directional (beam) antenna. Wire beams perform well but nothing beats a rotatable Yagi or quad antenna for real DX work. A portable 20 m QRP rig and simple wire antennas will work lots of DX from a campsite or motel room. Twenty is really a fun band.

15 AND 17 METERS

These two bands are very closely related in propagation characteristics. Both are daylight only bands under most conditions. Once we get above 20 m, sunspot activity becomes critical in producing band openings. There must be enough solar activity to yield a solar flux of at least 100 to 105 in order to produce band openings on 17 and 15 m. The higher the sunspot activity, the higher the solar flux. The higher the solar flux, the greater the ionization of the ionosphere and the farther the distances covered by signals on 17 through 10 m. Unfortunately when the flux is low (during the bottom of the solar cycle) there sometimes is not enough ionization present to cause a band opening on these higher frequency bands. Therefore, when you tune across 17 and 15 m (and even higher frequency bands) and hear no signals, it is a fair bet that the solar flux is not high enough to sustain sufficient ionization levels to produce a band opening.

Great bands for the QRPer, 15 and 17 m produce some outstanding DX openings into many areas of the world. They are ideal bands to use when working portable, since the overall size of the antennas are much smaller than for 20 or 40 m and this translates into less weight and bulk when backpacking in the bush.

12 AND 10 METERS

These two bands, like 17 and 15 m are daylight only bands that can produce some outstanding DX contacts during much of the solar cycle. These two bands offer some splendid DX during daylight hours when the solar flux is high enough to sustain a band opening. The bands form up several hours after local sunrise and by late morning they will be stabilized, yielding much DX. Summertime sporadic-E openings on 10 m will allow you to work stations within a few hundred miles. Normal short skip distance is about 900 miles. Using multihop propagation, signals on 12 and 10 m can circumnavigate the globe and give the QRPer

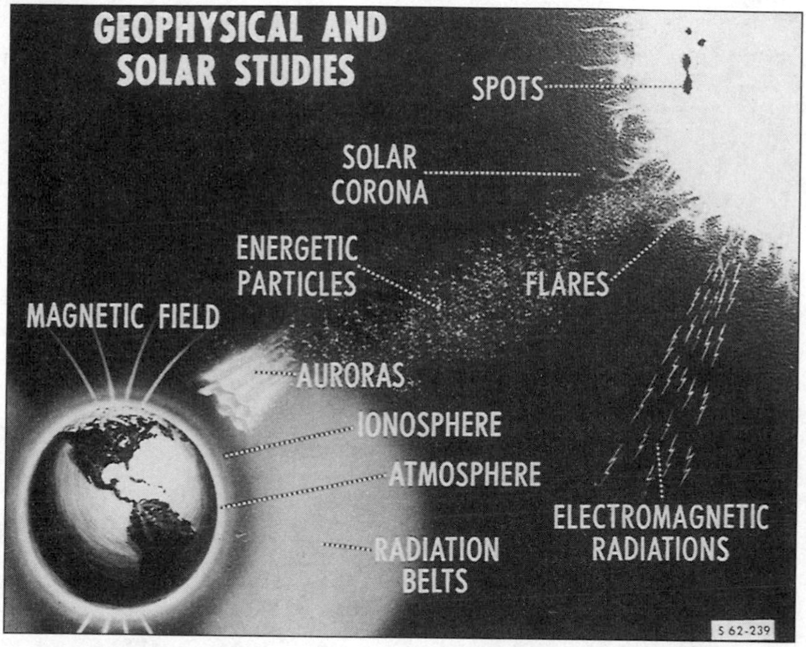

The NASA pictorial representation shows the influence of solar radiation upon the Earth's atmosphere.

quite a thrill. A monoband beam for either 12 or 10 m is quite small and can be turned using an inexpensive TV antenna rotator on a push-up mast.

WWV/WWVH PROPAGATION FORECASTS

Every hour, at 18 minutes past the top of the hour, WWV transmits HF propagation information, called Geophysical Alerts (or Geo-alerts), on 2.5, 5, 10, 15 and 20 MHz. (WWVH in Hawaii does this at 45 minutes past each hour.) This information can be used as a basis for planning QRP operations. **Figure 7-6** shows the WWV transmission schedule and **Figure 7-7** shows the schedule for WWVH.

Each hour the propagation forecast begins with the solar-terrestrial indices, a synopsis of the current propagation conditions, starting with the solar flux (indexed on the 10.7 cm solar flux emissions). Solar flux is a measure of the solar radio noise at the Earth's surface measured daily at 1700 hours UTC (Zulu) by the observatory in Ottawa, Canada. As the sunspot activity increases, the flux will increase and produce increased noise on the 10.7 cm band.

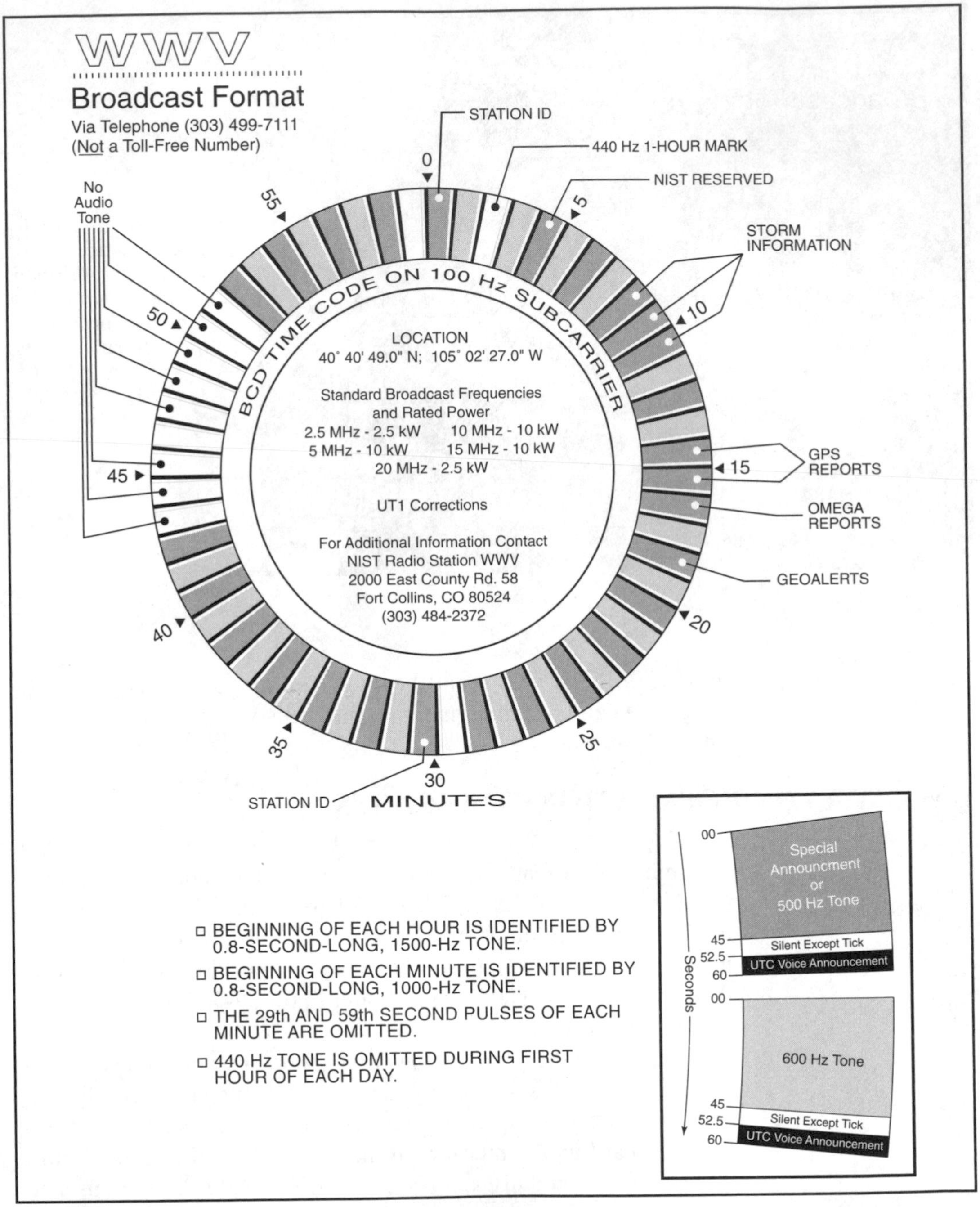

Figure 7-6 — This chart shows the hourly broadcast schedule for WWV, the National Institute of Science and Technology station in Fort Collins, Colorado.

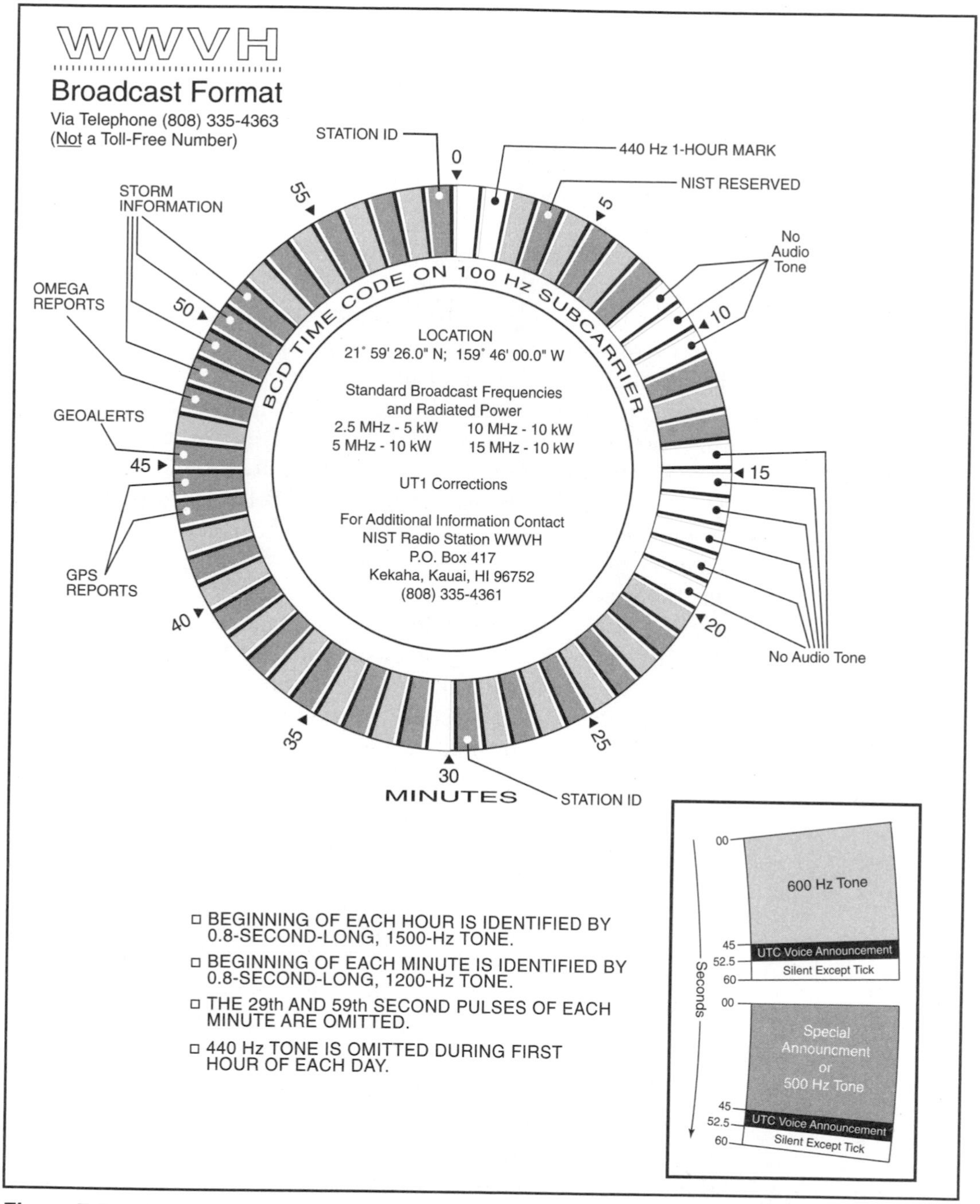

Figure 7-7 — This chart shows the hourly broadcast schedule for WWVH, the National Institute of Science and Technology station in Kekaha, Kauai, Hawaii.

Generally speaking, high solar flux values mean higher MUFs. The actual procedures for predicting the MUF at any given hour and for a specific path are quite complicated, though. Solar flux is not the sole determining factor. The angle of the Sun in relationship to the Earth, the season, time of day, exact location of the radio path and other conditions must all be factored into the equation. Long-term MUF predictions are even trickier since they involve even more variables.

After the solar flux reading comes the A index reading, which is a measure of the solar particle radiation, as determined by its magnetic effects. The A index represents a 24-hour average of readings taken every 3 hours throughout the day. The higher the number the greater the influx of solar particles. This equates to more absorption, weaker signals and greater fading.

Next the Boulder K index reading is given, which is measured every three hours and is a more accurate indication of the current influx of solar particles. In a nutshell, the lower the A and K numbers the more stable the ionosphere and the better the chances of working some DX. As the A and K indices increase, the more unstable the conditions in the ionosphere become and the less likely that reliable communications will be possible. During good conditions the A index will be between zero and eight and the K index will be no higher than two. During severe geomagnetic storms or strong X-ray events, the A index will jump to over 100 or higher, with an accompanying K index of seven or eight. When you monitor WWV/WWVH and hear high A and K indices, shut off the rig and watch some TV.

Following the reports on the solar-terrestrial indices, an overview of solar-terrestrial conditions for the last 24 hours is given. This explains what has happened over the previous 24 hours in terms of solar activity (disturbances on the Sun's surface) and the geomagnetic field.

Solar activity is basically transitory disturbances of the solar atmosphere as measured by X-ray emissions associated with solar flares. They are arranged into five ascending categories: very low, low, moderate, high and very high. An increase in X-ray emissions from the Sun's surface will cause an associated increase in solar atmospheric disturbances, ultimately resulting in degraded propagation here on Earth.

The geomagnetic field variations are based upon the amplitude of any disturbances on the surface of our Sun. These are clas-

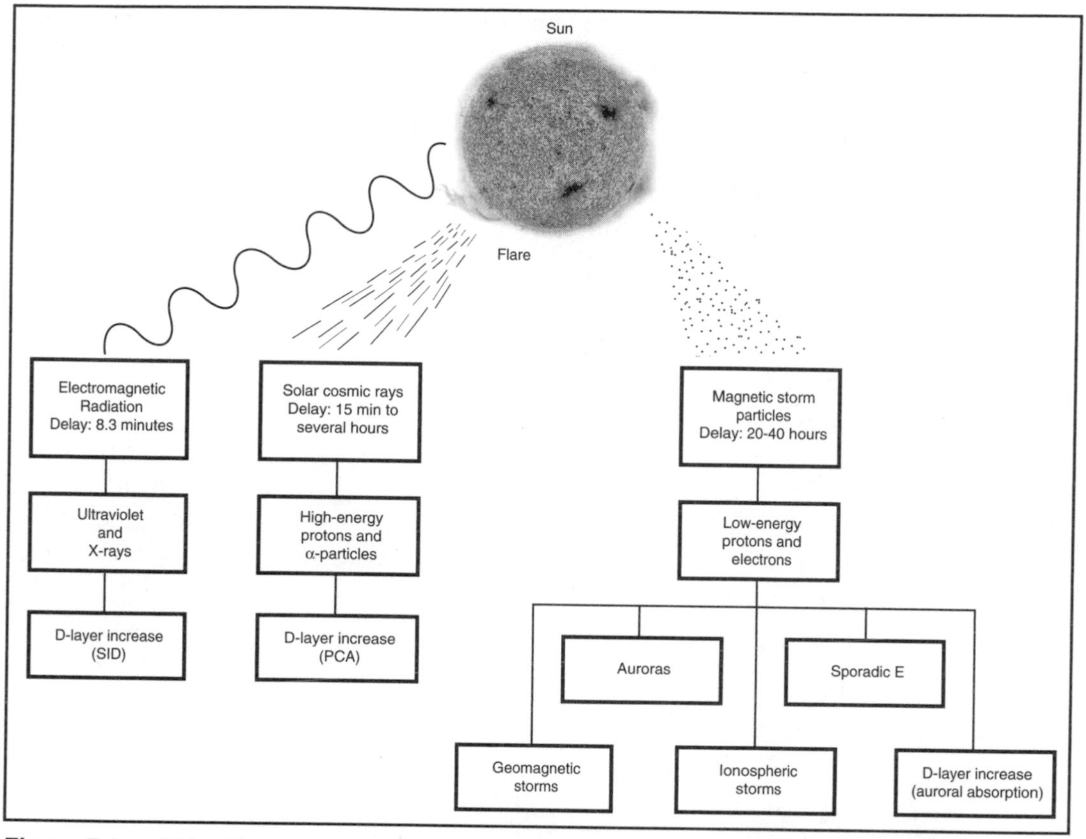

Figure 7-8 — This diagram shows the ionospheric affects of solar flare emissions. [Adapted from: George Jacobs, et al,* The New Shortwave Propagation Handbook, *by permission (CQ Communications, Inc, 1995), p 1-21.]

sified into six categories: quiet, unsettled, active, minor storm, major storm, and severe storm. Quiet is the most stable with severe storm being the most unstable. As solar flares and X-ray emissions increase on the Sun's surface, the geomagnetic field will become more disturbed and adversely affect propagation here on Earth. **Figure 7-8** illustrates some of the ionospheric effects resulting from solar flares.

Should there be an ongoing event on the Sun's surface, details of this event will be broadcast next. These events include major solar flares, proton flares, satellite level proton events, polar cap absorption and stratospheric warmings (stratwarm).

Virtually any time you hear one of these events taking place you can bet that propagation will be adversely affected.

Finally, at the end of each propagation broadcast, WWV/WWVH will include a prediction for the next 24 hours. This information is needed to assess whether or not conditions over the next 24 hours will be worthwhile and you can plan your QRP activities accordingly.

CHAPTER 8

QRP Station Accessories

Obtaining the necessary QRP radio and a useable antenna system are not all that you'll need to become effective in the QRP arena. What follows are some useful, if not indispensable, accessories that will be a great aid in assuring your station functions at peak efficiency and your QSOs are fun and rewarding.

QRP WATTMETERS

Unfortunately, your standard high-power wattmeter is ill suited for QRP use. Low-power accuracy is questionable at best. What you really need is a QRP wattmeter that will measure RF output signals down into the milliwatt region. There are several kits on the market. The Oak Hills Research WM-2 is a great one-evening kit that is not only accurate down to a few milliwatts, but also looks great. This meter is based on the Roy Lewallen, W7EL, design that appeared in *QST* a few years ago. Oak Hills Research upgraded the IC used in the original design, along with a few minor circuit adaptations and repackaged the unit into an attractive case with a very large meter movement. Calibration is done with a digital VOM and a dc power source. This well-crafted wattmeter has a very professional look and is accurate to within a couple percent of full scale. Speaking of scales, the WM-2 measures RF output from 0 to 10 W, 0 to1 W and 0 to 100 mW on three separate scales. The unit can run on an internal 9-V transistor radio battery or be powered externally by a "wall-wart" power supply. The WM-2 uses SO-239 UHF RF connectors for input and output connections. See **Figure 8-1**.

From time to time one of the regional clubs will offer a wattmeter kit (as well as other useful accessories) at very reasonable costs. Currently the North Georgia QPR Club (NoGA) offers a wattmeter kit. Check out their Web site for pricing and availability. (**www.nogaqrp.com**)

***Figure 8-1 — The Oak Hills Research WM-2 QRP Wattmeter makes a very nice beginning kit project for the first-time builder. This is a much-needed accessory around the QRP ham shack. The WM-2 is accurate down into the milliwatt range and is essential for keeping tabs on how your antenna and feed lines are working.* (K7SZ Photo)**

AUDIO FILTERS

With out a doubt, one of the most useful operational accessories (aside from a good quality keyer and paddle set) is an audio filter. Active audio filters have been around for over 30 years. My first experience with active audio filters was the MFJ CWF-2 back in 1974. I added one to my HW-7 and was amazed at the upgraded performance. One of my favorite mods on the HW-8 was to remove the two stage audio filter that came stock with the radio and add a CWF-2 in its place. Performance always perked up and I was able to pull the really weak stations out of the noise and QRM with only a slight outlay of cash.

Alas, the MFJ CWF-2 has gone the way of the Dodo Bird. There are, however, countless articles available on building active audio filters in the various ARRL publications. If you don't have the time or inkling to tinker, there are two filters available that come immediately to mind. First is the SCAF-1 switched capacitance audio filter from Idiom Press, (**www.idiompress.com**). This filter is a switched capacitance audio filter (SCAF) and is available as a kit or preassembled from the manufacturer. It features a 96 dB roll off per octave with a variable cutoff range of 450 Hz to 3.5 kHz. What does this really mean? This baby is sharp…very, *very* sharp, when it comes to the cut off frequency and the skirt selectivity. This means that you can snuggle in close to a QRO station or SW broadcaster, and still manage to pull the QSO out of the noise and garbage on the bands.

The SCAF-1 is a very user friendly accessory that essentially plugs into the headphone output of your radio and you, in turn, plug your headphones or speaker into the filter. Power is supplied from an external 12 V source. There is an "IN/OUT" switch on the front panel, along with a bandwidth control. Tune in your target station, switch in the filter and use the BW control to give you the best reception.

This filter is useable for CW, SSB, and digital modes. I have even used mine to augment my SW receiver while listening to SW broadcast stations. While the 3.5 kHz bandwidth is a bit narrow for good AM fidelity, it is definitely useable on SW, military and utility HF stations.

Ed Wetherhold, W3NQN, a long time contributor to ARRL publications, offers an improved version of his 88 mH toroidal inductor passive audio filter. See **Figure 8-2**. His new design uses potted inductors and a handfull of precision capacitors to achieve some rather astounding results. These filters are passive, meaning that there are no active devices included to amplify the incoming signal to overcome insertion losses. Actual losses using these newer potted filters are around 3 dB, which is easily overcome by slightly increasing your audio gain. Like the SCAF-1, Ed's filters go in-line with the audio output of the receiver or transceiver. The center frequency is fixed at around 700 Hz, but can be changed by carefully replacing certain capacitors inside the filter.

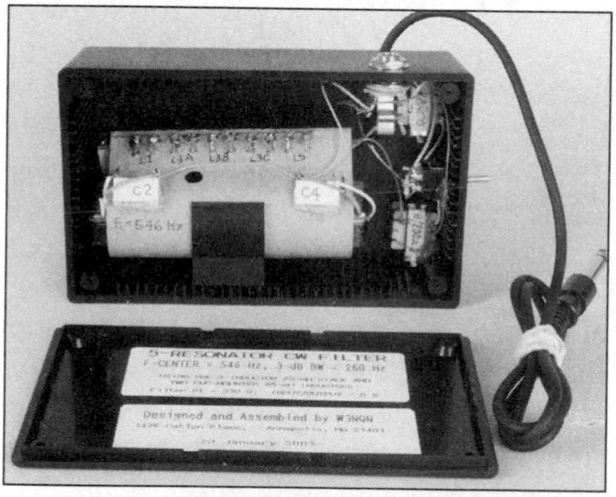

Figure 8-2 — This photo shows a passive CW audio filter built by Ed Wetherhold, W3NQN. The 2004 edition of **The ARRL Handbook for Radio Communicaitons** *includes construction details.*

The bandwidth of this new filter design is quite impressive, since the only parts involved are a series of L/C circuits. The overall effect of this filter is a definite narrowing of the receive audio with no ringing or "hollow" sounding audio that sometimes accompanies active audio filters. Using Ed's design, the adjective that comes immediately to mind is *mellow*. These passive filters are sweet.

One great application for these potted inductor passive filters is using them in conjunction with vintage (vacuum tube) equipment. Many of us "Boatanchor" enthusiasts are eager to use older, much less selective receiving gear on the ham bands. For AM phone operation you're not going to encounter many problems. CW operation is an entirely different matter, however. Seldom are the older receivers capable of the narrow bandwidths needed to provide good, clean audio for the CW operator. Even a well aligned Hallicrafters SX- whatever (the "X" designates crystal IF filtering) cannot be adjusted for good quality CW reception on today's bands. The overcrowded conditions and huge amounts of interference make for tough copy using older gear. Connecting one of Ed's new potted inductor filters, however, can really make a tremendous improvement in receiver performance, especially in the CW mode. For further details contact Ed Wetherhold, W3NQN at 1426 Catlyn Place, Annapolis, MD 21401-4208. Please include a self-addressed, stamped 9½ × 4 inch envelope with your request.

KEYERS AND PADDLES

Without a doubt, the finest CW keyer I have used to date is the LogiKey K3 available from Idiom Press, shown in **Figure 8-3**. The K3 provides for six memory banks, which can be chained for expanded replay. The K3 offers extended paddle input timing, which reduces errors and can help increase your speed. This keyer can be used with either positive or negative keyed rigs. This full featured keyer has incremented serial numbers for contests, and a whole host of memory functions including full beacon capability. The memory is non-volatile and is unaffected by loss of power to the keyer. The K3 is powered from an external 12 V source but consumes only picowatts when in the sleep mode. Styling matches the SCAF-1 audio filter also available from Idiom Press. If you want the ultimate contest keyer that comes in a very small package, I urge you to take a critical look at the LogiKey K3.

***Figure 8-3 — My favorite keyer: the Logikey K-3. This is a third generation of the Super CMOS keyer that was originally featured in the October 1981 issue of* QST*. Designed by Jeff Russell, KC0Q, and Bud Southard, N0II, the Super CMOS keyer also appeared in* The ARRL Handbook *for several years. This latest version has six programmable memories, runs on 9 to 15 V dc and has many programmable features.* (K7SZ Photo)**

Asking any CW operator about his/her paddle preferences is like asking a woman which pair of shoes she likes! There is a plethora of paddle sets on the market today. I have owned several including some from Vibroplex, Bencher, and Schurr. I like them all, but the most *comfortable* paddle set I have ever used is the NorCal paddle kit, later produced by Vibroplex as their "Code Warrior." This set of CW paddles uses magnets instead of springs, to provide a very positive tactile feedback to the operator. CW is effortless and ultra smooth. Two of my former students became journeyman machinists and collaborated to produce their version of the NorCal paddle set. This beautiful example of metal work uses five different metals, is rock stable and an absolute joy to use.

For rugged construction, you can't beat the Bencher BY-1 paddle set. My chrome BY-1 has gone everywhere with me, including trips into the bush for Field Day. They work and work and work and require almost no maintenance. They are robust, to say the least. Paddle tension and travel is easily adjusted by the set screws. The allen wrench is held beneath the paddle base for easy access. Bencher offers several versions of this paddle set, so check with them on prices and availability.

K7SZ Photo

My best advice on paddles is to try out as many as you can before plunking down your hard earned money (or plastic, as the case may be). Give them an honest try. Do a few Qs with the set(s) you are interested in to be sure that you fit them and they fit you. Like a well-worn shirt or broken-in boots, a good paddle set is comfortable to use for hours on end. Shop accordingly.

What we've covered here is just the tip of the iceberg when it comes to station accessories. MFJ Enterprises, and others offer hundreds of items of interest to the amateur radio operator. What we've highlighted in this chapter are several things that I feel are necessary to start fully enjoying your QRP efforts.

CHAPTER 9

Specialized QRP Modes

In this chapter we will cover some of the cutting-edge, high-tech aspects of the ham radio hobby as they relate to QRP. One of the most interesting things about QRP is its adaptability. While most of the QRP fraternity operates CW and phone, there are a growing number of QRP visionaries that forge ahead and try new modes with amazing success. Among these modes are digital (including RTTY, PACTOR, PSK31 and SSTV) and satellite communications, along with milli/microwatting, VHF+ and portable QRP operation. So, let's dive right in and explore some of these specialized QRP modes.

DIGITAL COMMUNICATIONS

Earlier I addressed the situation regarding the upsurge in newcomers to the HF bands. With the advent of amateur license restructuring by the FCC several years ago, thousands of hams, previously relegated to the VHF/UHF portion of the ham bands, now find themselves with HF privileges. Many of these hams are at a loss as to what to expect or how to properly conduct communications on the HF bands. They generally do not operate CW and are inexperienced with HF phone operation, so rather than explore their new territory, they gravitate to what they are comfortable with, namely 2-meter repeaters. In order to show these folks that there is life after 2 meters, and to provide an entryway to QRP, we are going to explore the digital world of ham radio.

Originally, the data communications portion of the first edition of *Low Power Communications* was only four paragraphs in length. Over the last several years, new digital modes and a plethora of new, inexpensive, easy-to-build equipment have risen to the forefront of ham radio. The combination of homebrew gear and cutting-edge digital technology has coalesced to offer the digital ham a new frontier for exploration. In this second edition of *Low Power Communications*, I have dedicated additional space to digital communications because it is the future of Amateur Radio.

Since the early 1990s, I have been involved with HF data communications using a multimode data controller (MMDC) and a QRP rig. My first station (a humble effort at best) consisted of an MFJ 1278 MMDC and a Ten-Tec Argonaut 509 transceiver for radio teletype (RTTY) and HF packet modes. It was a fun time, experimenting with these two modes and having keyboard-to-keyboard contacts with other digital hams. Success was virtually guaranteed as long as my signal strength at the receiving end was well above the noise level on the band. Although my early attempts involved a very primitive station, I had a lot of fun and gained some intimate and valuable experience with HF digital communications. One thing I immediately recognized was that the digital side of the radio hobby was where the future of Amateur Radio lay.

The real secret to HF digital communications at QRP power levels is maintaining a good signal-to-noise ratio at the distant end. Several methods are open to the low power communicator, not the least of which is a good-quality directional antenna. Any time you can concentrate your signal toward the other station you will improve the signal-to-noise ratio. Obviously, with the power limitation of only 5W, the digital QRPer must look to gain antennas to insure a good signal-to-noise ratio on the receiving end. The antenna chapter of this book will delve into this aspect in more detail. For now, suffice it to say that if you can erect a rotatable gain antenna of any sort, you'll be ahead of the digital game.

Digital Signal Processing (DSP), one of the new buzzwords in ham radio, is now offered on many commercial ham rigs. Using DSP in the receiver can greatly increase the signal-to-noise ratio. Once you convert the analog received signal to a digital signal, the DSP algorithms are able to separate and strip much of the noise from the intelligence. In some cases, DSP-processed signals can drag intelligence out of what seems like nothing but static. In short, DSP is about as close to real magic as you can get!

Another idea is to use data modes only when band conditions are prime. If the bands are in poor shape (severe fading, QSB, or noise, QRN), it's doubtful that your QRP signals will be heard (even with a directional antenna) with enough clarity to maintain solid communications. Watch the WWV/WWVH propagation forecasts and use the digital modes when propagation conditions improve.

Finally, try using the "error free" data modes when conditions are marginal. When your RTTY signals are being hit up pretty bad, a PACTOR, AMTOR or PSK 31 signal may not suffer the same problem. Mode selection *can* make a definite difference.

WHY CAN'T WE ALL JUST GET ALONG?

One pet peeve among QRPers is the appearance of data signals on or near the QRP calling frequencies, especially during QRP contests. If you engage in HF digital operations, please ensure that you abide by the established band plans. There is no need to venture out of the digital sub band. Likewise, even though 7040 and 14060 kHz are listed for both QRP and digital modes, be courteous. If you hear a lot of CW activity on or close to these frequencies, move up the band a bit. Chances are there is a QRP contest or operating event in progress, and weak-signal QRP operators won't appreciate your ratcheting digital signals. Most of the operating events are of short duration, lasting only 4 to 8 hours. The exceptions, of course, are the semi-annual QRP ARCI Spring and Fall QSO parties and several regional club contests that stretch over an entire weekend. The limited bandwidth of the digital modes, coupled with a good receiver and a courteous operator, will allow successful operation even under crowded band conditions. There's plenty of spectrum available so we can all coexist peacefully.

BANG FOR THE BUCK

Today's offerings of MMDCs are quite inexpensive when you take an objective look at the modes offered. Most, if not all, MMDCs offer CW transmit and receive capabilities along with standard Baudot teletype (RTTY) at 45, 75 and 100 WPM, ASCII teletype, H/VHF packet, AMTOR, PACTOR, and NAVTEXT modes. Most of these units cost under $300, which is a steal of a deal considering their versatility. Additionally, most offer optional DSP circuitry to improve readability under less than stellar band conditions.

While RTTY is fun, it is not 100% error free. AMTOR provides some level of error correction, but Packet and PACTOR protocols maintain 100% integrity of the transmitted message. PACTOR seems to be the most reliable with the best data throughput rates. I have talked with QRPers who regularly run PACTOR using 5 W and they have a ball! Nothing beats error-free traffic using QRP!

If you want detailed information regarding the current modes and operating trends in digital Amateur Radio, obtain a copy of the *ARRL's HF Digital Handbook* by Steve Ford, WB8IMY. Steve's book is a great read and presents the ins, outs and how-tos of HF digital operation.

BEHOLD! THE DIGITAL REVOLUTION IS AT HAND!

While CW and SSB modes are pretty well established on HF, newer, more exotic digital modes are rapidly catching on within QRP circles. Of course, RTTY (radio teletype) has been around for many years. And then there is the highly reliable AMTOR, and error-free PACTOR and Packet modes. However, PSK31 is the newest HF digital mode that has taken the ham radio hobby *and* QRP by storm. Over the last two years, PSK31, a digital mode pioneered by Peter Martinez, G3PLX, popularized in articles by Dave Benson, K1SWL; Howard "Skip" Teller, KH6TY; Steve Ford, WB8IMY; and George Heron, N2APB, has caused thousands of hams, world wide, to forgo the traditional communications modes in favor of digital HF communications. To say that PSK31 is the biggest innovation in ham radio since the inception of Packet and the Internet *combined* is an understatement. The popularity of this new digital mode is directly related to the simplicity of the relatively interference free, narrow band transmission system, ease of implementation and extremely low cost.

PSSSSST! WANNA BUY A RIG?

If you have a spare $45 (you read that correctly, $45) you might want to send it to Dave Benson, K1SWL, at Small Wonder Labs (**www.smallwonderlabs.com**), and pick up the PSK-80

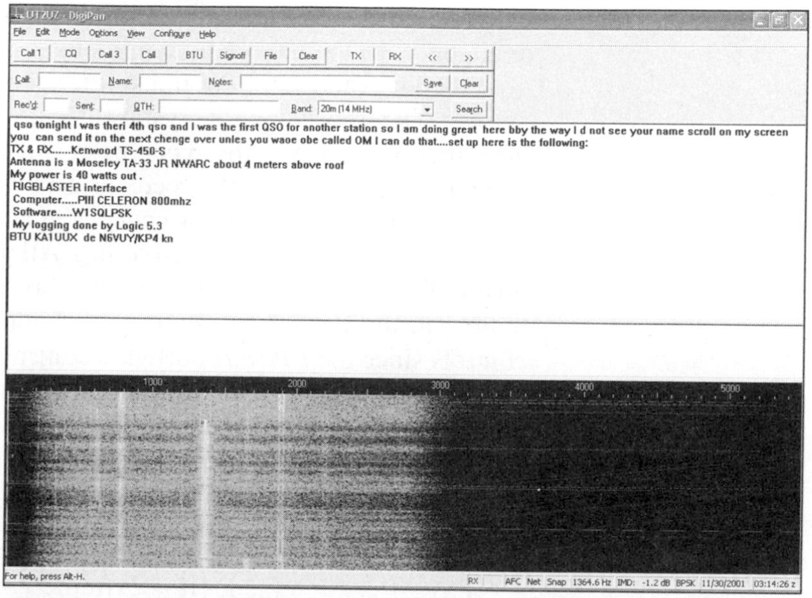

Working PSK31 QRP with* DigiPan *software. Each vertical line represents a PSK31 signal.

Warbler transceiver. This tiny kit is an entire 80-meter PSK31 transceiver you can build in a couple of evenings at the workbench. All you have to do to make it play is to obtain a piece of *free* software called *DigiPan* by downloading it from the Web at **www.members.home.com/hteller/digipan/**. Just hook up some cables between the transceiver and your computer's sound card. Add an antenna and power source and you're on the air with PSK31. It's *that* simple.

Why 80 meters? The availability of inexpensive TV colorburst crystals for the transceiver was the primary reason. Over the last couple of years, local 80 meter "Warbler Groups" have sprung up all over the US. This band provides very good coverage out to about 200 miles on a nightly basis. There is also the possibility of working some 80 meter "digital DX" when conditions are right.

The Warbler transceiver kit is the brainchild of Dave, K1SWL and George, N2APB. The idea was to provide radio amateurs with a simple PSK31 transceiver kit that would be easy to build and insure virtually trouble free operation. Since its debut at Pacificon in the Fall of 2000, the New Jersey QRP Club sold these kits like hotcakes. By mid-2002, Small Wonder Labs took over the production and marketing of these kits to insure their availability to the ham radio community.

The kit features a 5 W transmitter section and a direct conversion (DC) receiver with crystal filtering on the input. This feature greatly improves the receiver performance by providing a "tuned" stage ahead of the receiver mixer circuitry. This translates to a much higher performance standard than a stock DC receiver would be capable of delivering. All parts mount on one PC board. The kit comes with all parts plus the PC board. All the builder has to do is furnish a case and connectors. There are no controls since the *DigiPan* software controls all settings. The lure of operating digital QRP is the hook that grabs normal (?), low power communicators and turns them into digital mavens, lurking on 80 meters for PSK31 QRP QSOs!

GET BACK-TO-BASICS WITH DIGITAL QRP

PSK31 is a fun mode. It is extremely simple to operate. The free *DigiPan* software does all the work. The software offers the user a "waterfall" display on the computer screen covering several kilohertz of 80 meters. As signals are detected in the PSK-80's receiver passband, they are displayed as a set of small "railroad tracks" starting at the top of the display and trailing downward. All the QRPer has to do is place the cursor over the detected signal on the waterfall display, click the mouse button and the *DigiPan* software starts decoding the warbling tones! Wow! Just like a Tom Clancy movie! To conduct a QSO click the mouse on the **TX** button on the display and start typing. Simple, elegant and loads of fun!

The PSK31 signal bandwidth is extremely narrow, on the order of 60 Hz, so there can be many simultaneous QSOs taking place within a couple of kilohertz of spectrum. As you look at the *DigiPan* waterfall display, you can see several PSK signals being simultaneously displayed. It's easy to "read the mail" and find out who's on the air. Just click the mouse and watch the conversation scroll across the screen. PSK31 is *not* 100% error-free. However, many hams that have been in on the development of this mode from the beginning have the opinion that PSK31 is RTTY's replacement!

Neophytes and Old Timers alike are rediscovering ham radio with this potent combination of QRP and PSK31. Localized

Dave Hassler, K7CCC, snags digital QRP contacts from a portable station.

coverage provided by the 80-meter band combined with the ability to rag chew from the computer keyboard is a great marriage of innovative technology and traditional ham radio. This is "grassroots ham radio" at its very best.

Once you get your digital feet wet, try PSK31 on 20 meters using one of the Small Wonder Labs' PSK transceiver kits, or any other QRP SSB transceiver of your choice. The Small Wonder Labs' PSK-20 is a high quality digital transceiver that can open the world of HF digital DX to nearly everyone. This transceiver is slightly more complicated than the PSK-80 Warbler, but, if you take your time and follow the well-documented instructions, you will be rewarded with a digital 20-meter transceiver that fires up first time. The PSK-20 will allow you to work digital DX contacts all over the world with a transceiver you built yourself! We're talking *real* ham radio here! Homebrew rigs, DX QSO....*man!* This is living! The Small Wonder Labs' PSK20 and the NJ-QRP-Club's Warbler-80 are showcased in the equipment chapter, so drop by and have a look.

Have I managed to stimulate your imagination? I certainly hope so. There is much more to QRP HF ham radio than CW operation. There is a completely new digital QRP world out there that's up for grabs. Newly upgraded HF operators and seasoned QRPers can jump right in and be virtually assured of successful digital HF communications. Once you start making digital QSOs at the 3 or 4-W level using PSK31, you're in for a real thrill!

SATELLITE COMMUNICATIONS

The date was near the end of November 1983. The place: Newmarket, England. Darwin Piatt, G5CNP (now W9HZC) and I (G5CSU) were setting up our primitive "Earth station" in an attempt to contact Dr Owen Garriot, W5LFL, aboard the space shuttle *Columbia*, in this historic first attempt to contact a ham in space. Dar and I were using a homemade crossed-dipole antenna for 2 meters and a Heathkit HW-2036 10-W FM transceiver. Switching between the published up and downlink frequencies was done manually, since the 2036 could only handle ±600 kHz splits. Like I said, it was primitive.

Neither of us had any inkling of what Dr Garriot's downlink signal would sound like. We both figured it would be pretty "weak and watery" coming in from almost 300 miles out in space. Boy, were we wrong!

Of course, all the up/downlink frequencies had been published for months, but that didn't deter many of the UK hams from transmitting on the common downlink. The "Frequency Police" were out in force, trying to move everyone off the downlink.

We had computed the AOS (acquisition of signal) time and as the second hand drew near, both of us tensed. AOS came and went, no Owen Garriot. Lots of pandemonium on the frequency, but it was all terrestrial in nature. Finally, several minutes after the computed time we heard: "This is W5LFL aboard the space shuttle *Columbia*. CQ CQ from W5LFL!"

Dr Garriot's signal was so strong it almost bent the needle on the HW-2036 S meter. Man, what a signal! There was absolutely no mistaking what we'd heard. As soon as he quit transmitting, all hell broke loose on the downlink. Everyone, and I do mean *everyone*, was transmitting simultaneously on the downlink frequency! It was utter pandemonium.

Needless to say, with our puny 10 W we didn't manage a contact with W5LFL that day. We were, however, stunned by the clarity and strength of his signals here on Earth, especially since we were using a primitive setup. Dar and I recorded our reception times and ultimately we both qualified for a QSL card for receiving Dr Garriot's transmissions on this first "Ham in Space" event. What a thrill! It's something that both of us will remember until our dying day. This was my introduction to space communications.

The idea of communicating through an orbiting satellite produces images of *Buck Rogers in the 25th Century*. Satellite communications is intriguing, to say the least, and a challenge when done with QRP. Since the early 1960s, radio amateurs have participated in the development of space-based communications systems. Orbiting Satellites Carrying Amateur Radio (OSCAR) has been a mainstay of the Radio Amateur Satellite Corporation (AMSAT) for over four decades!

Currently there are whole crops of Low Earth Orbit (LEO) satellites (hereafter referred to as "birds") that can be accessed by a dedicated QRPer. Several are CW/SSB types that relay code or voice transmissions much like an orbiting repeater. There are several that are digital satellites that offer a "store and forward" messaging mode and others that offer 9600-baud packet transmission. There is even one LEO bird that has an FM voice transponder on board to facilitate FM voice contacts from orbit! And, let's not forget the International Space Station (ISS).

The ISS carries a ham station (ARISS) on board for use by the crew to contact schools on a scheduled basis, providing much needed public relations for the space program. Additionally, during rest periods, the ISS crew can use the station to make regular contacts with hams on Earth. When the dust settles and the entire ARISS station is finally operational onboard the ISS, there will be equipment included to operate from HF through the microwave bands with SSB, CW, FM, packet, ATV, compressed ATV and SSTV. Now *that* is impressive!

Prior to the Russian *MIR* space station's demise, cosmonauts and astronauts onboard *MIR* used 2-meter FM to contact Earth stations using both voice and 1200-baud packet modes. *MIR* also had UHF repeater (SAFEX II) aboard which gave many earthbound hams the chance to make terrestrial contacts via an FM orbital repeater. With as little as a 2W UHF hand-held transceiver and a 5-element UHF hand-held Yagi, it was possible to access the SAFEX II aboard *MIR*. Now that is real QRP!

MIR's VHF ham station became a critical communication

link after a *Progress* re-supply rocket missed its docking collar and slammed into a portion of the station causing a massive loss of internal power and a fire. Talk about a lifesaver—2-meter FM to the rescue!

With all these choices, it can become a bit intimidating to the uninitiated satellite communicator as to which satellite and which mode to try first. There is not enough room in this book to adequately cover the subject of satellite communications as it relates to low power communications. So, I will direct those interested in following this subject in depth to the excellent ARRL publication, *The Radio Amateur's Satellite Handbook* along with the *ARRL Satellite Anthologies*. Also, let's not forget AMSAT (**www.amsat.org**), who not only funds, builds and finds rides for Amateur Radio satellites, they also offer great software packages for tracking and working the various birds along with publications describing how to begin to enjoy satellite communications. If you are even remotely interested in satellite communications, I urge you to become a member and join AMSAT.

I can tell you from first hand experience that QRP and LEO satellite communications go hand in hand. What a thrill it is to work someone through an orbiting platform. With minimal gear and some simple antennas, you, too, can start enjoying the world of Amateur Radio satellite communications.

Unfortunately, newcomers to satellite communications are often scared off by pictures of huge steerable antenna systems and shacks full of specialized equipment. You do not need to spend tons of money on specialized rigs or antennas to start enjoying satellite communications.

The easiest way to start communicating with the LEO birds is to do it using gear you may already have on hand.

Launched on November 15, 1974, AMSAT-OSCAR 7 was equipped with a transponder that listened on 2 meters and retransmitted on 10 meters, as well as a transponder that listened on 432 MHz and retransmitted on 2 meters. It also carried beacons for 146, 435 and 2304 MHz, and a Codestore system of store-and-forward messaging. AO-7 has a unique place in satellite communications history, since it was the first satellite to be linked via the on-board transponders to AO-6, to provide a satellite-to-satellite relay between ground stations. This technique is commonly used today in military and communications satellites, but it was AO-6 and 7 that did it first!

Pat Gowan, G3IOR, operates his portable satellite station. (Photo courtesy of G3IOR)

In mid-1981 AO-7 mysteriously died. It is believed that an on-board battery failure caused by thermal stresses inside the spacecraft contributed to the premature death of this LEO bird.

Flash ahead 20 + years to 1992. Pat Gowen, G3IOR, in June of 2002, monitored the downlink beacon and telemetry of AO-7 from his home in Norfolk, England! Thought to be long dead for over 21 years, AO-7 suddenly came back to life. This bird is a usable comm platform as long as it stays in direct sunlight to produce electricity for the onboard systems via its solar panels. AO-7 resets itself each time it makes a darkness-to-light transition, so sometimes it fails to come on line. Although somewhat erratic, it is useable.

Talk about a comeback! Yes, you can access AO-7 if you can generate a CW or SSB signal on 2 meters while listening on 10 meters (Mode A). The satellite has a tendency to change modes unexpectedly, so you may also need the ability to transmit on 432 MHz and listen on 2 meters (Mode B). See the satellite frequencies listed in **Table 9-1**.

Gain antennas aren't a bad idea either, since their size is quite manageable at VHF/UHF. All you really need for Mode A is an FT-817 for 2 meter uplink and an HF receiver for the 10 meter downlink. Put a 2 meter downconverter on the HF receiver and move the FT-817 to 432 MHz to generate an uplink signal and you're good to go for Mode B.

OSCAR 7's orbit is polar, with an average altitude of about

Table 9-1
Popular Satellite Frequencies

Satellite	*Uplink Passband (MHz)*	*Downlink Passband (MHz)*
AO-7	145.850 to 145.950 432.125 to 432.175	29.400 to 29.500 145.925 to 145.975
Fuji-OSCAR 29	145.900 to 146.000	435.800 to 435.900
International Space Station	144.49 (FM voice/packet)	145.800
AO-27	145.850 (FM only)	436.795

1000 km or 620 miles. The LEO bird takes 105 minutes to completely circle the Earth, which means you will have between 4 and 11 "passes" per day depending upon your location. Each pass will last from one or two minutes for a pass close to the horizon to almost 15 minutes for a direct overhead pass.

Snag the latest Keplerian data (keps) from the AMSAT or ARRL Web sites, and update your satellite-tacking program with this latest orbital projection data. Then run a projection for the next couple of days and select a time that is optimal for you that matches up with a good near-overhead pass of AO-7.

On the appointed date and time, have your rigs tuned up and listen for activity in the downlink passband. Sometimes you may actually hear signals *before* the actual Acquisition of Signal (AOS), which is the time that the bird actually breaks over your local horizon. This is due to the propagation encountered on HF. As you track the pass, notice how signals seem to shift in frequency. This is due to the Doppler shift encountered when an object in orbit moves toward, then away, from a fixed point on Earth.

Slide up the band with your downlink receiver into the down-

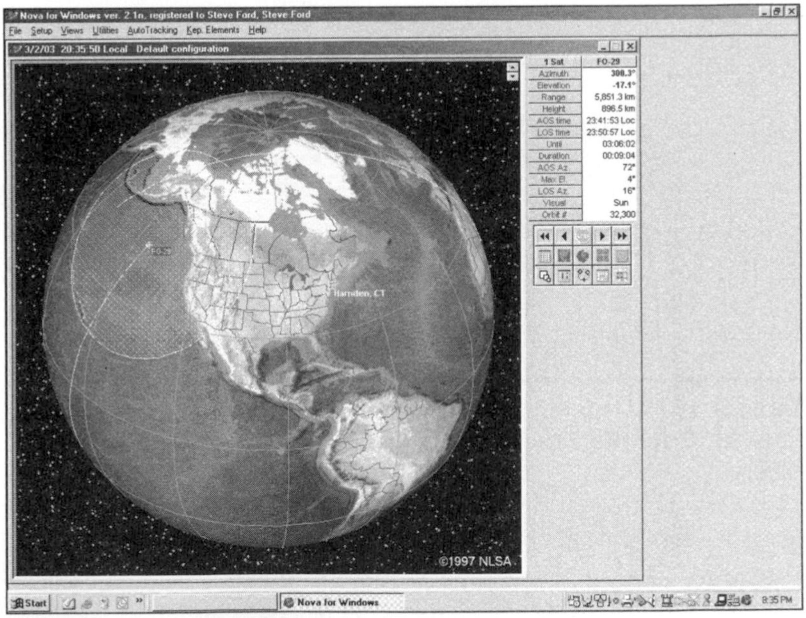

Nova *satellite-tracking software, available from AMSAT-NA at* *www.amsat.org*.

link passband of the transponder. Move your uplink transmitter into the uplink passband and fire off a string of "dits" on CW. If you are close to the proper frequencies both up and downlink, you should hear your signal coming back down to Earth. This lets you know that you are in the passband of the transponder.

At this time you can either call "CQ" or tune around for someone else calling "CQ" and answer them. I prefer the latter. Zero in on the other station's signal, move your transmitter to the proper portion of the uplink passband, shoot off a short string of "dits" to insure you're in the passband, then call the other station. You will notice that your signal and his signal both change as the satellite moves across the sky (Doppler shift, again). Once you make contact with the other station, *do not change your transmit frequency!* Track and compensate for the frequency change using your receiver. Otherwise, you will be chasing each other all over the transponder trying to maintain contact.

Joe Bottiglieri, AA1GW, makes an FM contact through the AO-27 satellite using an Arrow antenna and a dual-band hand-held transceiver.

Once you get a few LEO Qs under your belt, you'll be able to knock off three or four contacts on a long pass. However, when just beginning, consider yourself fortunate if you manage one or possibly two Qs per pass. If you decide to try using the satellites on Field Day, look for a short pass, low on the horizon, and use gain antennas. This will concentrate your signal down on the horizon, where the bird will be accessible. A low pass, preferably toward the East, if you're on the East coast, will limit the amount of other stations trying to access the transponder simultaneously with your QRP signals. I have successfully used this technique several times on various Field Days. You can learn more about AO-7 at this Web site: **www.experthams.net/ao7**.

As you can plainly see, there are many options for the aspiring QRP SATCOM operator. In addition to the birds discussed above there are several other LEOs worth taking a serious look at: Fuji-OSCAR 29 is a bird that listens for CW and SSB signals on 2 meters and repeats on 70 cm. AO-27 is an orbital FM repeater that is primarily active during evening passes. You can work this bird with a dual-band VHF/UHF FM transceiver. If you're running enough power (30 W or more), you may not even need a directional antenna. FM mobile operators have been known to show up on AO-27!

Although some might laugh at the idea of a QRP SATCOM station, I for one, know that working satellites using QRP power levels is doable. Unlike traditional QRP operation on the HF bands, when you venture into the realm of satellite communications you have a lot of new jargon to learn along with a new way of doing things. Don't let this deter you. If you are truly interested in satellite operations, grab some ARRL publications and log onto the various satellite communications Web sites on the Internet and go from there.

MILLI-MICROWATTING, OR HARD-CORE QRP

How low can you go? Now that is a good question. Since the late 1970s, QRPers have been taking up the challenge of milliwatting and microwatting in an effort to push the limits of their communications skills and station equipment. In 1989, Bob Moody, K7IRK and Bill Smith, WA6PYE, engaged in some milliwatt experiments on 10 meters using computer clock oscillator chips. With a power output of between 1 and 50 mW, they were able to make contact ova a 1300-mile path on a *regular* basis! On New Year's Day 1990, they completed a contact using only 4.62 microwatts for a total of 283,766,234 miles per watt!

In the spring of 1994, Bob Moody and Bill Brown, WB8ELK, shattered this 10-meter record by successfully using only 0.720 microwatts over a 1500-mile path for over 2 *billion* miles per watt! Since that time, milliwatters like W5ZPA have worked earned their Worked All Continents awards in just 10 days using only 40 mW of RF. Bob Moody, K7IRK, completed WAC using only 2 mW!

"Hard-Core" QRP, as it is called, is one of the areas of the low-power hobby that is attracting many QRPers. Attenuating the power output of your transmitter down into the milliwatt or microwatt range can provide the ultimate challenge.

While it can be argued that milli-microwatting on the 10-meter band is relatively easy because of the lack of noise and interference, doing the same thing on 40 meters, crowded with international broadcast stations, high noise levels and lots of interference, is something else again. Enter Fran Slavinski, KA3WTF, and Paul Stroud, AA4XX. Fran, located in Larksville, Pennsylvania, and Paul, located in Raleigh, North Carolina, had been actively pursuing the low-power record on 40 meters for several months in 1994. After some careful planning regarding the time of day best suited for the attempt, and locating a relatively interference-free frequency (7050 kHz), they were ready to attempt the record. On December 26, 1994, Paul fired up his solar-powered QRP rig, set his power level at 221 microwatts and started transmitting to Fran, 422 miles away in Pennsylvania. Paul included a "code word" that Fran had to copy correctly to ensure that the contact was legitimate. Contact was made, signal reports swapped, and the code word verified. Paul and Fran had broken the record and qualified for 1.9 million miles per watt!

A year later, they shattered their own record, this time using only 96 microwatts over the same 422-mile path. Fran successfully copied Paul's microwatt transmitter (and the code word) to raise the bar to 4 million miles per watt for 40 meters!

The publicity given these two events, coupled with the propagation upswing in the Solar Cycle, meant that others would jump onboard the Hard-Core QRP bandwagon. And jump they did. During the 1990s and into the new millennium, milli-microwatters were out challenging old band records. One spin-off of this facet of the hobby was the idea of beacon chasing. Here a QRP club or individual would activate a low-power transmitter on a given frequency and periodically change output power, normally going from around 1 W down into the microwatt range. All the while, the transmitter would be keying and each power level would have a unique code word, to validate reception reports. Beacon chasing still remains tops on the list of things to do for these Hard Core QRPers.

Milli-microwatting also has a following among the HF contesters. The semi-annual QRP QSO Parties, along with the milliwatt Field Day awards sponsored by the QRP ARCI, offer the Hard Core QRPer a chance to excel. For several years, I used 980 mW in QRP contests with success. It is amazing how well you can communicate using less than 1 W of RF power. Hard-Core QRP is an extremely rewarding experience. If you decide to take

Helpful Hints on Setting a Milli/Microwatt Record

1) Use a stepped attenuator to reduce your QRP transmitter output down into the milliwatt or microwatt region. (Editions of *The ARRL Handbook* from the mid 1980s to the mid 1990s show how to build a stepped attenuator.)

2) Use a nonreactive dummy load and an oscilloscope to verify your RF power output. Measure the peak-to-peak RF waveform across a dummy load, divide that ac voltage by 2 (to get peak voltage), then take that value times 0.707 to get RMS voltage. Square the RMS voltage and divide it by the resistance of the dummy load to obtain an accurate power measurement.

3) Find someone to work with who lives several hundred miles from your location. Plan, plan and plan some more to ensure success.

4) Spend time listening to the band you want to work, noting and recording propagation peaks at various times, congested and clear frequencies, and any other observations that might hinder or help your record attempt.

5) Erect the best antennas possible. For the 40 m record, Paul was using a 3-element wire beam at 60 feet, pointed north. Fran had a Radio Works Carolina Windom-II at 30 feet.

6) Location plays an important part. If your ambient noise levels are high, how are you going to hear very weak signals? Look for an RF quiet location where you have low band noise levels. Fran lives only 2 miles away from me but his location is so much quieter than mine, it is amazing. Outboard noise filters that actually null out noise sources are available and have been featured in *QST* articles.

7) Your current QRP rig should suffice provided it has a quiet receiver, excellent dynamic range, sensitivity and selectivity. Most of the current crop of rigs on the market meet these requirements.

8) Digital signal processing (DSP) can be quite useful when dealing with extremely low power signals. If your rig does not have DSP then you can add an external DSP filter unit. Several DSP programs exist that use your computer's sound card and specially developed software. These are extremely effective, in some instances giving you an additional 20 dB of signal to noise ratio.

9) If you have the real estate, loop and Beverage antennas are very low noise sky hooks you can employ to get the jump on noise.

up the milli/microwatt challenge, don't forget to document your efforts and share your experiences with the rest of the QRP fraternity on one of the Internet reflectors or a QRP club newsletter.

PORTABLE QRP OPERATING

When I look back over the last five years, I can hardly contain my excitement at the rapid, almost exponential growth of low-power communications. We QRPers have been treated to some outstanding and innovative portable transceiver designs, many of which are kit radios, from the fertile minds of engineers like Dave Benson, K1SWL, Wayne Burdick, N6KR, and Eric Swartz, WB6HHQ. The performance of these rigs is astounding, especially when you look at the parts count and simplicity of design.

The benefit of this abundance of QRP radio gear is the ease of which these rigs can be adapted to life on the trail. It seems that portable QRP operations have mushroomed over the last couple of years. Whereas the annual ARRL Field Day was *the* big to-the-field event in the recent past, now we have several major QRP contests that emphasize operation in the bush. Each month, the Adventure Radio Society (ARS) hosts a "Spartan Sprint" contest that is geared to micro-power and micro-weight. Penalties are incurred for excess weight of the station. Points are given for milliwatt/microwatt power levels and ultra-light rigs. The ARS also sponsors the annual Flight of the Bumblebees contest. The Eastern Pennsylvania QRP Club (EPA-QRP-C) sponsors the Appalachian Trail award for working stations on as well as operating from the A-Trail. The Northern California QRP Club (NorCal) sponsors their QRP-To-The-Field contest each year that is designed to fine-tune your Field Day skills prior to the actual event. There is no shortage of to-the-field operating events for QRPers.

What follows is an overview of portable QRP operation, the various types of gear and antennas used, and a sampling of operating events that will challenge you both physically and mentally.

ADVENTURE RADIO

Who says you can't take it with you? Certainly not anyone connected with the Adventure Radio Society. Born out of a desire to marry QRP ham radio with backpacking, hiking, camping and general outdoor actives, the Adventure Radio Society is a group of several hundred QRPers that advocate operating from the bush using minimal equipment. The ARS folks take QRP and raise it to an art form. Their quest for ultra-small and ultra-light equipment and accessories places this group of QRPers into a class by themselves.

AA4XX AND THE CAPE LOOKOUT LIGHTHOUSE

In June 2000, Paul Stroud, AA4XX, packed his microweight milliwatt station into a sea kayak and paddled out to the Cape Lookout Lighthouse on the Outer Banks of North Carolina to participate in the ARS Spartan Sprint. Paul's station consisted of a stripped down Small Wonder Labs SW-20+, microswitch CW key, one small earbud and a battery pack. Total weight of his station was 0.3 pounds, or roughly 5 ounces. Nominal power output from this miniscule rig is 100 mW with the ability to increase output to 200 mW, should conditions arise.

AA4XX's camp on the Shackleford Banks.

It took Paul several hours of paddling to reach the Cape Lookout Lighthouse from his departure point of Shell Point on Harker's Island. Upon arrival at the lighthouse, he erected his dome tent and antennas. Paul's antenna system consisted of a phased 2-element 20-meter vertical array, using half-wavelength 20-meter verticals spaced $^{1}/_{2}$ wavelength apart. Phasing between the two vertical elements was accomplished using a small homebrew phasing controller. The antennas were taped to a set of DK9SQ 10-meter telescoping fiberglass masts supported by PVC ground stakes that were screwed into the sand. The feed line consisted of RG-174 coax, lossy but quite light and more than adequate for use on 14 MHz. A parallel L/C matching network was used at the base of each antenna to match the high feed-point impedance to approximately 50 Ω.

The phasing controller provides instantaneous direction switching, which is a significant aid in eliminating interference from undesired directions while providing increased signal strength for the distant-end receiving station. A two-position switch on the controller provides either in-phase or 180° out-of-phase energy to each element. Thus, it is possible to select either "broadside" or "end-fire" patterns with this array. When you are only running 100 mW, every little bit of gain at the antenna helps.

The Spartan Sprint kicked off on schedule and AA4XX/4 found the 20-meter band in great condition. Paul started working other sprint stations with his microweight rig running only 100 mW of RF. Propagation and crowded band conditions weren't the only hazards that he had to contend with, however.

Those of you who've not been to the Outer Banks have never experienced "high tide" on a strip of sand that is only about 24 inches above sea level. The opening of Paul's tent was within 18 inches of the water at high tide! It's hard to concentrate on making contacts when you are keeping one eye on the oncoming water that relentlessly creeps toward the flap of your tent. Thankfully, the water started receding, and by 2245 EDT, Paul's tiny station was out of danger. About this time, KH6B appeared on frequency and Paul was able to complete a contact with Hawaii using only 200mW of power!

Paul managed to win the "skinny" division of the June 2000 Spartan Sprint. Not only did he have a lot of fun combining ham radio and his kayaking hobbies, but the trip out and back from the Cape Lookout Lighthouse was filled with some wondrous sights. His photograph of a herd of wild ponies grazing on the grass at the edge of the Shackleford Banks was marvelous!

WA3WSJ QRV FROM MT WASHINGTON

Ed Breneiser, WA3WSJ, has this "thing" about the Appalachian Trail. Ed believes in the Zen of ham radio. He and his Elecraft K2 transceiver are one with each other and one with the Appalachian Trail. Don't ask me when all this started; even Ed doesn't know. All that I really know is that it's hard to keep Ed and his rig off the A-Trail for any length of time.

One sunny Saturday in October 2000, I met Ed in Northeastern Pennsylvania, near Route 309, where the A-Trail crosses the highway. We had decided to operate from the trail's nearby parking area (well within the 1000-foot rule outlined in the Appalachian Trail Award sponsored by the Eastern Pennsylvania QRP Club). Ed's station consisted of a Elecraft K2 and a multiband dipole fed with twinlead. My station was a Red Hot NorCal-20 transceiver and a 20-meter dipole. We had a lot of fun for several hours and managed quite a few A-Trail contacts.

Later, Ed voiced his desire to hike a portion of the A-Trail in New Hampshire, culminating with a trip to the top of Mt Washington, the highest peak in the Northeast. Ed's goal was to operate from the peak of Mt Washington and help those hams needing New Hampshire contacts for their A-Trail Awards.

In mid-July 2001, Ed and his daughter flew to New Hampshire and began hiking the A-Trail. They managed to climb Mt Washington and operate from the summit, capping (sorry about the pun) a wonderful vacation, which included hiking, camping, ham radio and some quality time with his daughter.

SPEAKING OF THE A-TRAIL...

One of the hottest new awards QRPers are chasing today is the *Appalachian Trail Award* sponsored by the Eastern Pennsylvania QRP Club. The brainchild of Ron Polityka, WB3AAL, this is a multifaceted award that has a little something for everyone. The basic award is presented to anyone who makes one QRP contact with another station in all 14 states that the Appalachian Trail runs through. Then there is the award for a QRPer who works all 14 Appalachian Trail states while actually operating from the trail. And the list goes on.

This award has spawned a new generation of QRPer who wants to take to the bush and operate from an historic site, like the A-Trail, Route 66, etc. With the proliferation of quality QRP portable rigs on today's market, the active QRPer can have a lot of fun and enjoy the out of doors simultaneously.

TRAIL FRIENDLY ANTENNAS

Wire antennas tend to dominate the portable QRP scene. Why? Because they are ultra simple to build and deploy. A quarter-wavelength end-fed wire and matching counterpoise terminated into an "L" tuner, while not the most efficient portable antenna, will work quite well in most instances. There is one critical thing to keep in mind when selecting/building a portable antenna for trail use: many times it will be impossible to erect the antenna at heights above 20 feet due to dense foliage, trees or lack of suitable support. Therefore, I suggest you shelve the idea of using full-sized dipoles at heights above 25 to 30 feet.

Ed Breneiser, WA3WSJ, who spends lots of time on the A-Trail, has found a workable solution to the portable multi-band antenna problem. Ed built a dipole antenna he calls the *A-Trail Dipole*. This antenna consists of a 40-meter dipole (33 feet per leg), which is fed with 300 Ω twinlead or ladder line transmission line. Using a tuner with a 4:1 balun to terminate the feed line before it goes into the transmitter allows Ed to work all the HF bands except 17 meters. Full details on the A-Trail Dipole are found in the antenna chapter of this book.

There are several versions of portable vertical antennas available on the market for trail use, too. Ventenna markets a nifty multiband HF portable vertical called the *HFp*, that, when disassembled, fits into a small case only 14 inches long, which is great for the backpack. The HFp offers the portable QRPer a 40 to 10 meter vertical antenna, support guy wires, guy stakes and radial counterpoise in a compact package. Add 25 feet of RG-174 coax and you're ready to roll. Ventenna also sells optional add-on kits for 80 meters. Performance is about what can be expected of a ground-mounted, electrically shortened vertical antenna. In my own case I have managed my share of DX with the HFp over the last couple of years that I have operated QRP from the bush.

James Bennett, KA5VGS, designed and built the *PAC-12* antenna that features separate coils for the HF bands from 60 through 10-meters. This antenna can be ground mounted or can sit atop a tripod for above-ground mounting with elevated radials. A full account of how to construct your own homebrew version of the PAC-12 antenna is contained on the New Jersey QRP Club's "Website on CD ROM" available from the club for $10. Check out their Web site for ordering info at: **www.njqrp.org**. If you don't feel like building this antenna from scratch, James offers a kit version. This antenna is a very good performer, considering it cost me less than $25 to build my original copy of James' design. While not quite a center loaded design, this HF vertical has a lot going for it, including cost and overall size when disassembled for transport. Additionally, the PAC-12 won the HFPack "An-

AD2A checks out a PAC-12 portable antenna. **(photo courtesy NJQRP)**

The Elecraft KX1, an ultra-portable QRP CW transceiver that's ideal for use in the great outdoors.

tenna Shoot Out" at Pacificon QRP gathering in 2002. James' PAC-12 bested all comers. This "Shoot Out" has become a regular feature of Pacificon. The tests are conducted on a test range using a dipole reference antenna. All antennas (both commercial offerings and homebrew antennas) under test are compared to this reference and are rated in descending order as to the amount of dB below the reference. The PAC-12 stole the show! Not bad for a homebrew antenna design that you can duplicate for about the same amount of money it takes to fill the gas tank on your vehicle!

VHF+ AND QRP

When we think about traditional low power VHF+ operating, images of a radio ham with his trusty hand-held transceiver come immediately to mind. QRP VHF+ operation is routinely done using FM and repeaters. The repeater, obviously is used to boost the QRP signals in order to cover larger areas than could be done with simplex operation. In this case, QRP gets you into the repeater, but the repeater does the actual work.

In reality, "weak signal" VHF+ operation normally starts with 50 W and goes up to full legal limit in order to overcome path losses encountered at VHF and higher. SSB is the preferred mode, although you'll encounter CW from time to time. A combination of high-gain antennas and high RF power levels typifies the VHF+ "weak signal" station. Ironic, no?

Having said all this, however, there is still a place for "real QRP" on the VHF/UHF bands. Many QRPers think that VHF and QRP are mutually exclusive terms. Nothing could be further from the truth. With the in-troduction of the multi-band/multimode FT-817 transceiver by Vertex Standard (Yaesu) several years ago, VHF+ operation using QRP has skyrocketed. The FT-817, with its VHF+ coverage (6 meters, 2 meters and 70 cm) on SSB, CW and FM is uniquely suited to hilltopping, hiking/backpacking, contest roving and emergency communications. Many primitive areas through out the United States encourage hiking hams to monitor 146.52 MHz (FM) for emergency wilderness traffic.

Okay, I have a confession to make. I have a love affair with my FT-817. So much so, that for several months my wife actually thought I had a mistress! No, I don't have stock in the company, and I paid full price for my unit along with the optional accessories. I just *really like* my FT-817.

Bob Witte, KØNR, is ready to hike up Mt Herman with his QRP VHF/UHF station. **(WA6TTY photo.)**

"BECAUSE IT'S THERE..."

Want to have some real QRP fun? Grab your VHF+ rig (FT-817 or whatever you have handy) and scoot up to the top of a large hill or mountain. Include a portable 4 or 5-element 2-meter beam along with a similar 70-cm antenna and you are in for some great fun hilltopping using VHF/UHF. You don't need to drag along a lot of battery power for just a few hours on a hilltop: just the rig, antennas, grid square map and a way to log your contacts.

Combine hilltopping with a major VHF/UHF contest, add a "rare" grid square, and you'll be the hit of the party! It is truly amazing how far you can work using QRP power levels on VHF+ from an elevated location. Hilltopping requires no special skills, just the desire to have some fun by combining a love for the outdoors with VHF+ ham radio. Although the FT-817 is my choice for a hilltopping rig, older VHF+ QRP rigs like the ICOM IC-202, 402 and 502 radios will work nicely, although the power output will be only around 2W. Yaesu's earlier offerings: the FT-690R (6 meters), FT-290R (2 meters), and FT-490R (70 cm), fall into the category of "luggable" as opposed to "hand-held" rigs. These radios weren't true QRP rigs either, running 10 W output. The Japanese company, Mizuho, also marketed the MX-6 (6 meter) and MX-2 (2 meter) handheld transceivers that will work in this application.

My personal choice for portable antennas are those from Arrow Antennas. Allen Lowe, NØIMW, of Arrow Antennas (**www.arrowantennas.com**) has an outstanding selection of portable antennas for VHF/UHF along with a host of fox hunting and Search and Rescue (SAR) accessories. Allen's antennas have elements that disassemble and store inside the boom. While I use separate 2 meter and 70 cm Yagis, Arrow Antennas also offer a 144/432 MHz combo (3 elements on 2 meters and 7 elements on 70 cm) that assembles on a single boom and uses only one feed line.

Of course, you could always opt to build your own VHF/UHF antennas. A quick trip through the *ARRL Antenna Book* and the *ARRL's VHF/UHF Antenna Classics* will give you plenty of ideas on easy-to-build skyhooks that will enhance your QRP signal. Recycling old TV/FM antennas is one source of scrap aluminum for your antenna project. Conversely, you could go to any of the large home improvement stores and search their stores of aluminum tubing and angle stock for the raw materials for your antennas.

One final thought about QRP and VHF+: why not combine hilltopping, VHF+ and satellite communication? Imagine hiking up a hill, butte or mountain, setting up your gear and contacting the International Space Station or making some FM contacts through AO-27! It doesn't get much better than that!

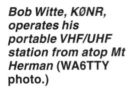

***Bob Witte, KØNR, operates his portable VHF/UHF station from atop Mt Herman* (WA6TTY photo.)**

CHAPTER 10

Emergency Communication and QRP

First came the bombing of the World Trade Center Towers in 1993 by an organized terror group. Then came the horrific September 11, 2001 attacks by terrorists on the Pentagon and, once again, on the World Trade Center Towers. The 1993 incident should have been the proverbial wake up call, but it went unheeded for the most part. It took the events of September 11, 2001 to shock us out of our innocence. That day Americans suddenly realized that terror attacks were not something that happened to other people in Third World countries. Thankfully, the Amateur Radio community was willing and ready to step up and provide emergency disaster relief communications on very short notice when Public Service/EMS repeater systems and trunked radio systems failed, and the much-lauded cell phone infrastructure totally collapsed.

Jump ahead almost two years. The East Coast Blackout of August 14, 2003 was yet another red flag to anyone involved with emergency communications and Homeland Security. Millions of people across the Northeast, Midwest and Canada were affected in the biggest power outage in the history of the US. The much-touted cellular infrastructure once again failed miserably. Public Service/EMS repeater systems and trunked radio systems were severely degraded or totally out. Amateur Radio operators, once again, saved the day.

These events underscore the need to become involved with emergency communications. If you have not joined a local

The events of September 11, 2001 shocked us out of our innocence.

ARES and/or RACES group yet, please do so. Don't forget to log onto the ARRL Web site (**www.arrl.org**) and sign up for the basic emergency communications course. Remember, if you're not properly trained, you are useless in an emergency.

Emergency communications and QRP are not mutually exclusive terms. Over the years, I have come to rely upon my QRP gear to double as my emergency communications equipment. Good antennas and proper operating procedures are the order of the day. It isn't so much the amount and type of equipment you deploy with, but how you effectively use what you have.

My own emergency planning revolves around something I call my "Jump Kit," which is nothing more than the bag (or bags) you grab when you walk out the door to perform emergency communications (EMCOMM) duties. While I have had some people tell me, point blank, that QRP has no place in EMCOMM, I, and many others, totally disagree. QRPers have top-notch operating skills, and the size and portability of our gear makes us a valuable EMCOMM asset. While many people *think* they need at least 100 W of RF power to maintain communications, QRPers know differently—and we practice what we preach.

ANATOMY OF A JUMP KIT

Your Jump Kit should contain virtually everything you need to enable you to set up and operate a communications station during an emergency. Planning and assembling a Jump Kit is no easy task due to the variables encountered. The following information is provided as a guideline to help you assemble your own Jump Kit, which, by the way, is *never* completed. It is a dynamic entity that changes as your emergency communications requirements change.

The ARES guidelines want the emergency communicator to be self sufficient for a minimum of three days. A more realistic goal is 10-14 days. This means that you need to assemble lots of "stuff" besides the obvious radio equipment. This includes several changes of clothing, a toilet kit, spare glasses, any medications you regularly take, wet-weather and/or cold-weather gear along with high-energy food items when there is no food available at your site. Don't forget a small tool kit, spare batteries, antennas/feed lines and RF/audio adaptors, log sheets, ARRL message forms, pens/pencils, emergency lighting, maps, compass/GPS unit, whistle, first aid kit, sun block, insect repellent (100% DEET), weather radio and/or scanner. Sometimes you'll need to provide your own tent for shelter along with a sleeping bag. Now do you get the picture? The challenge is to reduce this huge mass of equipment to a manageable amount. In emergency situations, mobility counts!

THE LISTS

Your first task should be to make some lists, separating them into several categories: **Radio Gear, Station Accessories, Power, Antennas, Personal Items, First Aid/Meds** and **Shelter/Food**. List absolutely everything that you might need during an extended stay in the bush. Once you're done, take a close look at your lists and think about the things you can *absolutely do without* and then remake your lists. Once you've done this several times you'll likely be down to a list of essential items that will allow you to perform your EMCOMM duties at an acceptable level for an extended period in the field.

Now is the time to assemble everything you have on your various lists and take a hard look at what you now have to pack and haul to a deployment location. Believe me, you will be amazed at the sheer amount of gear that you have listed to take with you in your Jump Kit.

Go back to your lists. This is the time to get real about *exactly* what you need to deploy with in order to do the job. Chop it up! That's right, get out the red pen and start lining through items that you can do without. You'll be surprised what you can leave out of your Jump Kit and still get the job done.

When you're done, once again assemble the equipment on the various lists and take stock of the size of the pile. If you're still not satisfied, go through the exercise of paring the lists again. Eventually, you will get to a point where you will strike a balance between what you need, what you "think" you need and what you *have to take* into the bush.

Once you've finalized your lists they need to be laminated and included in the Jump Kit since they provide an accurate inventory of all your gear for both pre and post emergency deployment.

Having done all the previous planning, selecting and paring, you need to assemble your entire Jump Kit and hit the road, literally. Pack it all up and go for a weekend camping trip or take it on a business trip. The idea is to get it all together and find out if it all works and whether or not your plans are in keeping with your communications needs. Nothing beats an occasional "shakedown" deployment just to see if the system works. If you're ever called on to provide emergency communication, these times of pre-disaster deployment evaluation will be time well spent because you will know exactly what you need, how to set it up and operate it effectively. In a disaster scenario, seconds count.

WHICH RADIO?

The type of RF gear you'll need will be dictated by the requirements of your local ARES/RACES group and the agency(ies) they serve. The majority of my emergency communications revolves around VHF FM voice and packet. Therefore, I need a good VHF rig, some portable gain antennas, and possibly a packet TNC, laptop computer and printer.

HF is nice to have and the pressure is on throughout the Commonwealth of Pennsylvania to have all county Emergency Operations Centers (EOCs) active on HF SSB. Should the individual

Figure 10-1—My current Jump Kit. Spare clothing, foot stuffs, meds, etc are carried in a small suitcase that stays in the truck.

ARES/RACES member have portable HF SSB capability? Absolutely! It's another method of communications that might prove crucial in an emergency.

My Jump Kit contains a Yaesu FT-817 transceiver (HF as well as 6 and 2 meters and 70 cm). It can be argued that 5 W might be a little on the light side for reliable HF voice communication, which is why I also include a homebrew 40-W HF amplifier (**www.hfpack.com**) and a 35-W VHF/UHF amplifier, a Mirage BD-35 (**www.mfjenterprises.com**). Alternatively, you could opt for the new Yaesu FT-857 transceiver, which has the same frequency/mode coverage as the FT-817 but provides up to 100 W output on HF and 6 meters, 50 W on 2 meters and 20 W on 70 cm. These power levels are continuously variable upwards from 5 W, so you can adjust the output to provide a balance between reliable communications and available power budget. I also include a VHF/UHF hand-held transceiver (a RadioShack HT-420) and a 200-channel scanner, just to round out the mix. **Figure 10-1** shows my current Jump Kit. Spare clothing, foot stuffs, meds, etc are carried in a small suitcase that stays in the truck.

IT'S IN THE BAG!

You'll need a carrier for your gear. Use what you want with an eye toward mobility. Over the years, I have used brief cases, duffle bags, a "Street Bag" from a police supply house, a plastic trunk, a toolbox and a soft-sided cooler. Each carrier worked fine at a given time but was replaced when my comm requirements changed. Ruggedness counts, too.

Currently my EMCOMM radio gear lives in a soft-sided cooler while my clothes, toilet articles and other supplies live in a small carry-on-size suitcase with retractable handle and casters on the bottom for ease of mobility. In my truck, I also carry a small two-man dome tent, ground cloth, rain fly, all-season sleeping bag with foam padding and *chemsticks* for emergency lighting. Also packed in the bed of the truck is a four-element 2-meter beam antenna, several lengths of RG-8X coaxial cable and a 10-meter collapsible fiberglass mast, along with several sections of steel pole that telescope together to form a 15-foot-tall antenna mast.

My Nissan Frontier 4-wheel-drive pickup (with a crew cab) is also fully outfitted with HF-2 meter communication gear, an 11-meter radio, a 200-channel VHF/UHF scanner, GPS and a 300-W, 12 to 110 V dc-to-ac power inverter to provide limited ac power from the truck's battery. I felt that the fully outfitted mobile communication capabilities of the truck was warranted after the 1996 flood of the Susquehanna River that forced our evacuation of south Wilkes-Barre and the subsequent loss of my entire ham station at the house.

Your emergency communications requirements may be more or less intense than mine. The advice in this chapter is provided as a starting point to help you conceive and assemble you own Jump Kit so you can become a more effective emergency communicator. The name of the game is readiness. You have to temper your desire to cover all the bases with the realization that you can't have everything you want during emergencies. Plan your Jump Kit carefully, get the necessary training and you'll become a valuable EMCOMM asset.

CHAPTER 11

Vintage Radio and QRP

There is something extremely romantic about using vintage vacuum tube equipment on today's ham bands. Tube gear has a certain ambiance: it glows in the dark! You get a warm fuzzy feeling when operating vacuum tube gear that is decidedly lacking in the transistorized, digital radio equipment we use today. Tube equipment transports the operator back to a time when CW was fun, AM phone was king and SSB was the new kid on the block.

Many of us "Old Timers," who started our ham careers back in the late 1950s and early 1960s, cut our teeth on vacuum tube equipment. For us it's a walk down memory lane when we fire up a vintage ham station or homebrew transmitter. Others within our hobby gravitated toward tube technology as a way to enrich their ham radio experience: to broaden their understanding of our fantastic hobby, if you will. For whatever reason we choose to play the vintage vacuum tube radio game, it all boils down to fun. There is also a certain challenge involved, for these older receivers are not nearly as selective as the state-of-the-art rigs we currently use. Crystal filters employed in vacuum tube receivers, while considered more than adequate for the bands of the '50s and '60s, are woefully inadequate under the crowded band conditions that exist today. Therefore, we have to sharpen our listening skills in order to use this older gear with any hope of success.

Restoring vintage gear has special significance for me along with many others who collect, restore and use these priceless old rigs. First of all, when you take the time to restore a classic receiver or transmitter from the vacuum tube era, you are preserving a valuable piece of radio and electronics history. At one time some of this vintage gear was the absolute top of the line equipment and commanded a small fortune in 1940s and 1950s dollars. These rigs were the cutting edge of radio technology of their era. And amazingly, they work quite well when restored.

When these older rigs are gone, they will be gone forever. I view my own humble efforts in vintage gear restoration as a legacy that I can leave to a future generation of radio aficionados. I love old gear, even the higher powered, non-QRP equipment, and get a real thrill when I see an old rig come to life after sitting abandoned for years, enduring decades of neglect. To watch an old vacuum tube receiver light up and produce sounds from the "ether" is absolutely amazing to me.

One of my favorite old receivers is the National NC-57. Mine came from Ed Wetherhold, W3NQN, in Maryland. Ed upgraded his station and gave me his NC-57 complete with the receipt from when he bought it in 1947 in Alaska! After a noteworthy restoration, which included many new tubes along with replacing virtually all the capacitors *and* resistors in the set, I not only have a nice piece of historic radio gear, I also have quite a bit of the history behind this particular unit.

If you start procuring and using vintage ham gear, be forewarned: You *will* get sucked into the "Boatanchor Vortex" and your involvement with ham radio will change forever! Hey, it's a good thing!

In the first edition of *Low Power Communication* I presented my quest for a vintage QRP station that culminated in a Hallicrafters S-53A receiver and a HT-18 exciter/transmitter. I have included this saga in the 2nd edition, primarily because it's a great success story.

RESTORING A VINTAGE QRP STATION

Over the years I have restored and used several complete vintage Hallicrafters and Drake stations along with a host of odd receivers and transmitters. One day I started thinking, why not marry QRP and Boatanchors?

At this point I started an intense search for some tube type QRP gear of the 1950s or '60s. Cost was definitely a factor. I did not want to spend a small fortune for my Boatanchor QRP station. Since I specialize in restoring Hallicrafters gear, I figured that was a likely place to start. Surely Bill Halligan had marketed

Figure 11-1 — Featured here is my vintage QRP pair. The Hallicrafters S-53 superhet receiver and HT-18 exciter/VFO make a cute little pair of mini-Boatanchors. The receiver is quite broad, so some sort of audio filtering is a must. The exciter works in the CW and FM modes (no AM or SSB, sorry) and provides a clean signal on the air. Despite their age and lack of sophistication (compared to today's rigs) this little "glowbug" QRP station can provide hours of entertainment and fun on the bands. Not only do you experience what it was like to operate on the HF bands 50 years ago by restoring older tube-type gear, you are preserving a valuable portion of our ham radio heritage. **(K7SZ Photo.)**

a low power exciter or VFO that I could pair up with my Hallicrafters S-38. After searching Chuck Dachis' book, *Radios by Hallicrafters*, I found what I thought would be a perfect vacuum tube QRP station: the S-53A receiver and the HT-18 VFO/exciter.

The S-53A is an eight-tube general-coverage receiver manufactured from 1950 to 1958, covering 550 kHz to 31 MHz and 48 to 54.5 MHz in five bands. AM, SSB and CW reception is possible. Cosmetically it looks like a cross between an S-38 and an S-38D on steroids! The HT-18 is a seven tube, 4 W VFO/exciter produced from 1947 to 1949, that features CW and NBFM output on 80 to 10 meters, and shares the S-53 look. The pair was designed as an entry level receiver/VFO-exciter combination. See **Figure 11-1**.

Availability was one of my primary concerns. With an eight year production run of the S-53A I figured that there was an excellent chance of quickly finding a receiver. The HT-18 transmitter, with its two year production run, would be another matter.

Where does one find vintage gear? There are several sources but the most productive one I've found is the Internet. The BA Swap List (**baswap-list@foothill.net**) along with the Boatanchors List (**boatanchors@listserv.tempe.gov**) are two *free* subscription reflectors that cater to vacuum tube enthusiasts who want to swap information, ask questions and buy/sell/trade Boatanchor gear.

I listed a want to buy (WTB) for the S-53A/HT-18 combo and within two hours I had two offers to sell me HT-18s! So much for hard to find! Shortly after that, two S-53As became available via Internet e-mail. I opted to buy both S-53As and HT-18s, since it never hurts to have a spare rig around just to compare wiring, check voltages and have a source of spare parts.

One HT-18 did not work. All the tubes lit up but I could not get any RF output. This rig quickly became the parts rig for the other HT-18. Both the S-53As worked but needed a good cleaning and complete recapping. A quick look under the chassis revealed 18 old wax paper capacitors that needed to be replaced as part of the restoration process for each receiver. I like to use "Orange Drop" molded capacitors to recap my Boatanchor gear. These and other restoration supplies are available through *QST* advertisers.

While waiting for the capacitors to arrive, I checked all the vacuum tubes on a transconductance tube tester. I replaced weak or substandard tubes with new, old stock (NOS) tubes. There was some backlash in the bandspread tuning, so I restrung the main and bandspread dials using new dial cord and springs. While the dial cords and subfront panel were removed, I cleaned the entire chassis and lubricated the capacitor tuning shafts.

Proceed with caution as you begin the cosmetic cleanup of the case, front panel and chassis. Stay away from aggressive cleaning agents as they can attack the silk-screened lettering, which is almost impossible to replace. In the instance of one of the S-53As, using only mild dish detergent and warm water, I almost destroyed the "Hallicrafters" logo! This is the first time I had seen this happen on a Hallicrafters radio.

Do not use ammonia-based window cleaners on dial glass, as the cleaner will totally eradicate the lettering on most dials! I use diluted dish detergent and/or Murphy's Oil Soap for most cleanup chores. If something is really stained and Murphy's won't cut it, I then resort to Simple Green diluted 1:2 with water. Avoid gritty, cleanser-type cleaners at all costs. Don't forget to thoroughly wash the metal parts with clean water to remove all residue from the cleaning agents.

I like to use a liquid brass polish, which has a very mild scouring action, to improve the luster of the cabinet and front panel. Used sparingly, the brass polish will restore the original color and luster of the case.

After thoroughly wiping the brass polish off the metal surfaces of the case and front panels, I applied a good grade of silicone automobile wax to seal the paint, brighten the metal surfaces and counteract the effects of finger oils on metal. This final polishing results in a like-new appearance for the old rig.

Recapping an old Boatanchor is probably the most tedious task in a restoration. It's done one capacitor at a time by desoldering the old capacitor leads then routing the new capacitor leads to their respective tie points and resoldering the new cap. In many instances, it is a good idea to include some insulation over the new capacitor leads to keep things from shorting to other parts of the chassis or circuit.

A trip to the test bench to align the rig was in order to improve dial calibration once I replaced the wax paper capacitors. The excellent instructions in the manual make this phase of the restoration a snap. Both S-53s aligned quickly using a Hewlett-Packard 606D RF generator and an HP 5245L counter to spot the generator's frequency. Similarly, I had no problems aligning the HT-18.

After the clean-up and alignment, it was time to put the two units together and see if my master plan for vintage QRP would actually work!

One *big* problem was dial accuracy on both the receiver and the exciter. Most of us have become very used to digital readouts on our transceivers. Going back to the analog world takes some adjustment. Add to this the inherent dial inaccuracies of the entry-level receivers of the 1950s where the width of the dial pointer can cover 5 kc … well, you get the picture.

To counteract the dial calibration problem, I built a small crystal oscillator that used some HC-25U crystals for 3560 kHz, 7040 kHz and 14060 kHz, the 80, 40 and 20 m QRP calling frequencies. See **Figure 11-2**. Use this crystal oscillator in the same manner as a crystal calibrator. Turn it on and zero beat the receiver to the crystal oscillator's frequency on the proper band. Then zero beat the exciter against the receiver's frequency. With

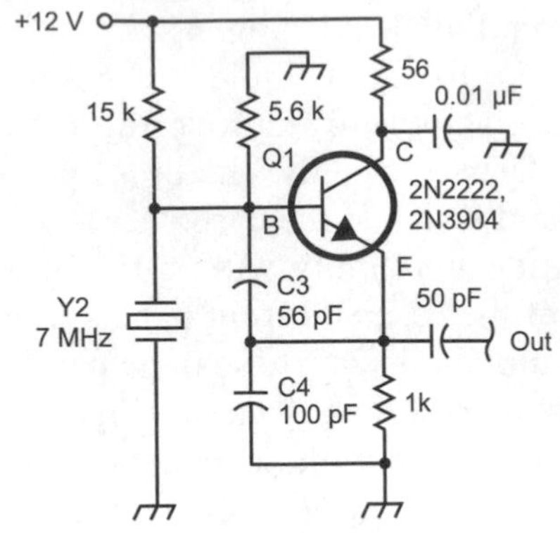

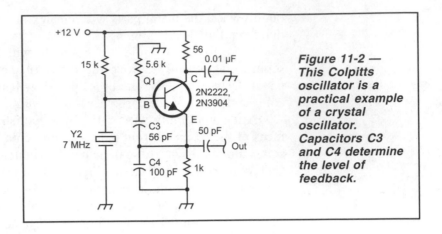

Figure 11-2 — This Colpitts oscillator is a practical example of a crystal oscillator. Capacitors C3 and C4 determine the level of feedback.

a little practice, this procedure becomes second nature, just like it did for hams in the 1950s.

I had two major concerns regarding the transmitter: stability and signal purity. After many on-the-air contacts and some careful listening to my own transmitted signal on a separate receiver, I was convinced that Hallicrafters did an outstanding job of taming the HT-18's output. There was some drift after an initial 30 minute warm up, but this is normal for vacuum tube equipment.

Well, does it work? Ooooh, yeaaaaaah! I was absolutely amazed with this vintage QRP rig's on-the-air performance. Listening to a very simple receiver without all the modern bells and whistles takes some getting used to, but it's a lot of fun.

Life with a glowbug (the QRP version of a vacuum tube rig) is interesting, to say the least. Transmit/receive switching is accomplished manually, via toggle switches on the front panels of both the HT-18 and S53A. QSK it's not!

Dial inaccuracies aside, entry level vacuum tube receivers have a very wide IF bandwidth. This means you are going to hear several CW QSOs in progress simultaneously. Here is where an outboard active audio filter or DSP filter unit is very beneficial. I found that the RadioShack DSP filter works very well when plugged into the headphone output of the S-53A. See **Figure 11-3**. If you want to stay in character with the 1950s motif, try constructing a passive audio CW filter using 88 mH toroidal inductors. Editions of *The ARRL Handbook* from the early '80s through 1994 have the details. The 2004 edition of *The ARRL*

Figure 11-3 — Want some portable DSP? In my opinion, this is the best bang for the buck. Unfortunately, RadioShack no longer markets this little device. The good news is that you can find these little outboard filter boxes at ham radio flea markets for about $20! They make great add-on filters for QRP rigs, and they also provide a built-in amplified speaker. I have used these devices with solid-state rigs as well as vacuum-tube Boatanchor receivers with excellent results. **(K7SZ Photo.)**

Handbook even has an updated design by Ed Wetherhold, W3NQN. See Chapter 8, QRP Station Accessories in this book for a photo and more information about such filters.

Finally, stability is not a strong suit in low-end tube-type ham radio gear. Drift is a way of life. Mechanical stability of the receiver leaves a lot to be desired and any sudden stress on the main tuning or bandspread dials can cause the frequency to shift several kilohertz! Ambient air currents also drastically affect the receiver stability. While it is manageable, it can result in constant attention being given to the receiver and transmitter frequencies during a QSO. You will definitely get the flavor of the '50s when you use this glowbug gear on today's bands.

Over the first few weeks of use I managed to work some stateside and European DX on 40 and 20 m using my vintage QRP station. Contacts using the S-53A/HT-18 are exciting, and since I am using a station that is 50 years old, I never have a problem finding something to throw into the conversation to make it more interesting.

I learned a lot from this experience. Performance is truly amazing for such simple circuitry. The Hallicrafters company designed and produced some of the world's most affordable radio equipment. Emphasis was placed on performance and moderate cost, and Hallicrafters' success is self evident.

If your Amateur Radio life is "ho-hum" and you feel you are up for something completely different, why not try vintage QRP. Not only will you reacquaint yourself with what ham radio life was like 40 or 50 years ago, you'll have a lot of fun in the process and you will be helping to preserve a portion of our hobby's history.

DO HAMS STILL REALLY USE TUBE GEAR?

While preparing a vintage radio segment for my *QST* "QRP Power" column, I posted a message on several Internet reflectors looking for hams that had built and were using tube type gear. I was deluged with replies and pictures! Unfortunately, I did not have room in that column for even one third of the pictures and captions that I received. Therefore, I have decided to include some of the rigs that QRPers have made and currently use on the air in this section of the book. My goal is simple: start building! The parts exist to construct a complete vacuum-tube QRP station. It might take a little scrounging, but any QRPer worth his or her salt can certainly find the necessary parts, including vacuum tubes, to build a vintage low power station.

Not feeling up to building a superhet receiver from scratch? Find a vintage ham band receiver like a Drake 2B, Hallicrafters SX-117, National NC-303, Collins 75S3-B, or similar old receiver, and restore it to its original glory. Use that fine vintage receiver in conjunction with a home brew transmitter of your choosing for some great QRP fun plus a lot of nostalgia.

Need help finding circuits to duplicate or use as a spring board for your own designs? The older editions of the ARRL's *Radio Amateur's Handbook* from the 1950s and '60s contain a wealth of information and schematics of tube gear. Hamfest flea markets are a great place to scrounge for parts, vintage vacuum tube receivers in need of restoration, Novice transmitters of the era, etc. There are probably one or more tube aficionados in your area that will be more than happy to Elmer you through the process and help you design and build a vintage QRP station. So have at it! Don't let fear hold you back; jump right in and give vacuum tube gear a try.

The *Glowbugs* Internet reflector is an excellent place to frequent. On that list reside some of the most helpful people I have ever encountered when it comes to vacuum tube gear. The Glowbugs Web site (**http://s88932719.onlinehome.us/glowbugs.htm**) has information and a link to sign up for this outstanding reflector dedicated to vintage gear and vacuum tube QRP transmitters.

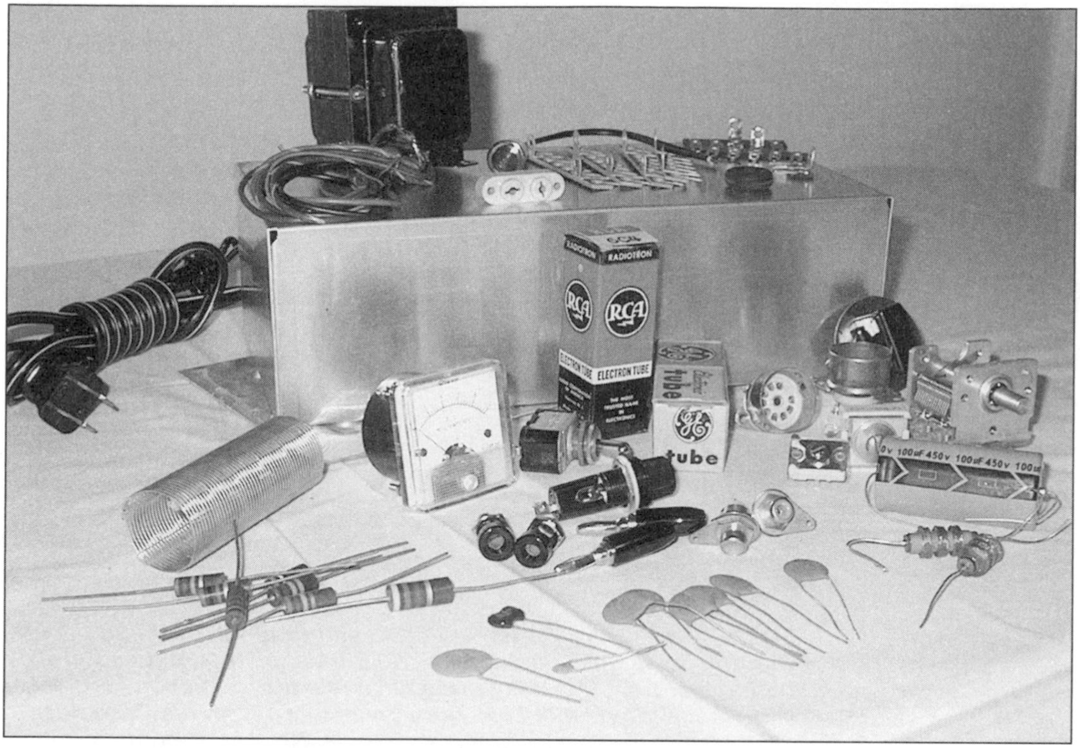

(A)

(B)

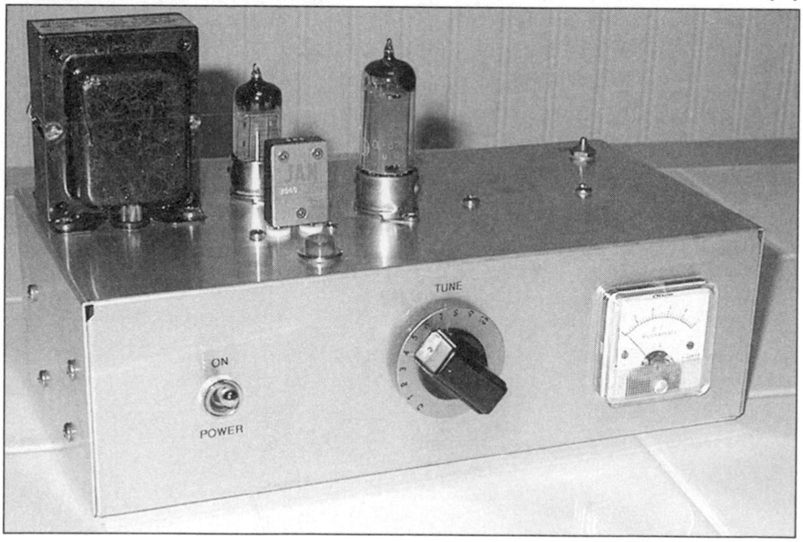

Figure 11-4 — Harold Kraus, K2UD, started with an aluminum chassis and a few parts, and built this two-tube "Simple Transmitter for the Beginner." The project first appeared in the October 1968 issue of QST, by Don Mix, then W1TS. It features a 6C4 as the crystal oscillator driving a 5763 PA tube. Power output is about 5 W on either 80 or 40 meters. **(K2UD Photos.)**

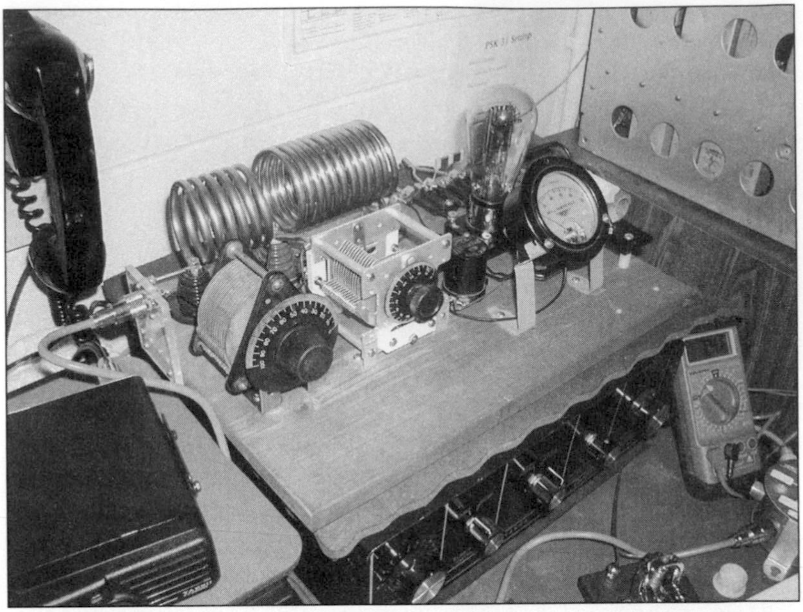

Figure 11-5 — Gary Carter, WA4IAM, shows off his building talents with this beautiful example of a TNT (tuned plate, not tuned grid) transmitter from the 1920s. This is Gary's first homebrew project, and what a job he did! This circuit was taken from the December 1929 issue of **QST** ***and features a single 210 tube yielding a whopping 2 W output. Gary made a few changes in the original design, including shunt feeding the plate circuit. He also stayed true to form using square buss wire, rubber coated wire and as many "period vintage parts" as possible. Congratulations, Gary, on a beautiful rig.*** **(WA4IAM Photo.)**

One word of caution: the voltages used in vacuum tube gear are *lethal*! Use extreme caution when building, operating or troubleshooting vacuum tube equipment. Work with one hand in your pocket to help protect you from unexpected electrical shocks that would otherwise travel through your heart and cause arrhythmia or possibly death. Always check your test leads for frayed or broken insulation that could spell disaster while troubleshooting. Above all, always, *ALWAYS* be aware of what you are working on. Even circuits that have been powered down sometimes can "bite" you due to capacitors failing to discharge regardless of the fact that they are equipped with bleeder resistors. This is especially true when working with really high volt-

***Figure 11-6 — Jim Riff, K7SC, from Scottsdale, Arizona, constructed this nicely done 160-meter Hartley AM/CW transmitter. The rig features a 6L6 final amp and a 6K6 used as the modulator. A 6SN7 is used as the mic preamp. High level modulation in the plate circuit is achieved via a choke in the B+ line to the PA. Power output is up to 8 W. By using separate coils, Jim can operate 160 through 10 meters on AM and CW. Nice job, Jim.* (K7SC Photo.)**

age power supplies like those found in medium power transmitters and high power linear (ugh!) amplifiers.

In addition to the National NC-57 receiver, my current vintage station consists of a Heathkit HR-10 receiver (circa 1961) and an E.F. Johnson Viking Adventurer Novice transmitter. The Adventurer is crystal controlled and has a single 807 PA which yields about 50 W input power and about 20 to 25 W output. T/R switching is done via a double pole double throw ac relay that is operated by a toggle switch. The relay changes over the antenna from the receiver to the transmitter (or vice versa) and mutes the receiver during transmit. Not even close to full break-in (QSK) but definitely just like I did it when I was a Novice, for this station is a duplicate of my original Novice station. While the transmitter is a bit ragged cosmetically, the HR-10 makes up for this in "pretty" and performance.

Other "Boatanchor" (BA) gear that lives at K7SZ consists of a Drake 2B receiver with the matching 2BQ speaker/Q-Multiplier accessory. This receiver is probably one of the best CW receivers of the vacuum tube era. With selectable IF filtering down to 500 Hz, coupled with the razor sharp performance of the Q-Multiplier, CW signals seem to pop out of a very quite band. The performance of this receiver is quite startling. The 2B is relatively small compared to other receivers of the day. It is well designed and can hold its own, even on today's crowded bands.

My other BA receiver is a classic Hallicrafters SX-117 (circa

(A)

(B)

Figure 11-7 — Eddy Swynar, VE3XZ, built this version of the "QST Super 12," a 12-tube receiver. The original design is from an article by Ross A. Hull in the March 1929 issue of* QST*: "Improving Short-Wave Phone Reception. A Modern Super-Heterodyne for Short-Wave Phone, Code and General Broadcast." Eddy was inspired to build this version by T. J. Lindsay's article in the February 2002 issue of* The Old Timer's Bulletin. *(VE3XZ photos.)

1961 to 1964) that came my way from my good friend Mike Zane, N6ZW (ex K6URI), out in Lodi, California. Mike had been having troubles with some instability on 10 and 15 m, and sent the SX-117 to me saying he'd give it to me if I could fix it. Several of the tubes checked questionably on the transconductance tube tester so I replaced them. The problem persisted, so I very carefully applied some DeOxit cleaner/lubricant to each tube socket and reseated each tube several times to clean off any corrosion that might have built up in the intervening 40+ years. There were seven original "Black Beauty" capacitors still in the receiver and I replaced them using Sprague "Orange Drop" 5% capacitors. Those Black Beauty caps are notorious for failing and should be replaced any time they are encountered in a piece of vintage gear. Next I performed two complete alignments, following the manual to the letter. The instability problem was gone and the receiver functioned great with an MDS in excess of 0.25 microvolts on each band! Not bad for a 42-year-old receiver! During the 2003 CQ World Wide phone contest I copied all continents on 10 meters with no instability at all.

The SX-117 has special historical significance for me since that was the receiver (along with the matching HT-44 transmitter and HT-45 Loudenboomer kilowatt linear amp) that I used at the

Figure 11-8 — Steven Johnston, WD8DAS, built this tube version of the Tuna Tin 2. **(WD8DAS photo.)**

WA7CDH club station at Yakima Valley College in 1964. I always wanted to duplicate that particular station since it was the first real high-end ham station I had used in my short Amateur Radio career up to that point. Now, if I can only find the HT-44 (and, yes, the HT-45) along with the HA-5 "TO Keyer" I will have achieved nirvana!

The last section of this chapter includes photos of some outstanding homebrew QRP gear that has been constructed by our fellow QRPers. Remember everything you see here is in use on the bands. Many of the designs are from the 1920s and feature "period" parts from that era. As you can see from the photos, these are all works of art and the QRPers who built them are to be commended for their attention to detail, outstanding craftsmanship and their dedication to the craft.

Epilog

The following is text that appeared in a past *QRP Power* column in *QST*. This column provided the most feedback of any column I wrote over the entire four-year period of doing the QRP column in *QST*. Obviously, it touched on a "hot topic" and the feedback I received was overwhelmingly in favor of what I had written. Therefore, I felt that this information about my views on QRP would make good reading for anyone not having access to the original article.

QRP PHILOSOPHY

Often we, as QRPers, take our hobby a little too seriously. If someone else's idea of how to pursue QRP is not the same as our own, we get defensive and confrontational. I have seen this happen on the QRP reflectors, as well as in person. The first thing we must understand is that if you gathered 100 QRPers together in one room and asked them to write down their idea of what QRP was and how they planned to pursue the hobby you would assuredly get 100 separate answers to the question.

DIVERSITY

The idea that we all have a slightly different take on our hobby is a good thing. It offers the possibility of expanding one's horizons, when interacting with other QRPers. Left to our own devices we would stagnate and our hobby would grind almost to a halt.

QRP is considered by mainstream Amateur Radio as a group of arcane individuals who practice the esoteric art of CW. We all know that this is a rather narrow vision of our facet of the hobby. In fact, QRP is the fastest growing area of ham radio. With the advent of low-cost, high-performance kit radios, thousands have flocked to the ranks of low-power communication and become homebrew aficionados in the process. The recent "digital rush," thanks to PSK31, has opened the HF bands to many newly licensed hams. Until now, many of these hams have had little interest in our side of the hobby, fearing that a lack of CW proficiency would prove fatal to their dreams of enjoying QRP.

New rigs like the Yaesu FT-817 and ICOM FT-703 Plus have opened the door to portable QRP operation for thousands (if sales records are accurate). The idea of using a tiny rig as a "pedestrian mobile" station has captured the imagination of the folks who hang out on the HFPack reflector. Many innovative ideas are discussed (and cussed) on this reflector, resulting in a very lively discussion group and an outstanding source of information on portable HF operation.

Our diversity is our strength. By encompassing many different methods of enjoying QRP, we evolve and grow. This evolutionary process is vital to our survivability if we are to attract new blood into the hobby.

WHAT IS QRP?

Strictly defined, QRP is the pursuit of ham radio at the five-watt level: nothing more, nothing less. While for some, QRP has become a lifestyle bordering on a religion, it's still just a hobby. Many times non-QRP hams have been treated to a litany of reasons as to why QRP is "right" and high power (QRO) operation is "wrong." Man, nothing turns people off faster than having someone get up in their face and collectively tell them off for doing what they are having fun doing. It's like debating who's having more fun fishing: the guy with the ultra light tackle or the dude with his trusty Garcia pole and 12-pound-test line. Both are having fun. Neither is "wrong."

This concept extends even into the ranks of QRP itself. There are those who practice the "minimalist concept," where everything is homebrew, and all the antennas are wires at low height. I have seen, and in one instance, been part of a discussion on an Internet group regarding minimalist radio. Typically, the minimalist says that true QRP can only be accomplished by using the bare minimum of gear, usually homebrew, and wire antennas. Power for the radio is obtained via solar charging or the small generator attached to the side of the hamster cage. Big beams, log periodics, Rhombics, etc. are evil and not in the "True Spirit of QRP."

Of course, the counter argument is that QRP is a power level *only*, and efficient antennas level the playing field. Minimalist QRPers view all others as being heretics while non-minimalists view their counterparts as being totally out of step with reality and the concept of QRP.

There is no "right" or "wrong" here. However you choose to pursue QRP is your business. If it is fun and you enjoy your time on the air using the minimalist philosophy, that's great! If, on the other hand, you really can't get into the minimalist movement, then that's okay, too, as long as you're having fun doing QRP your way. Either way, you are a better operator because you're using minimum power to communicate on a global basis.

WHEN QRP ISN'T ENOUGH

I know I'm going to take some heat for this topic, but it must be said. All grandiose, righteous, ostentatious motivation aside, there are those times when five watts isn't going to work. I hate to be the one to tell some of you this, but it is true. In those instances there is nothing wrong or shameful about going to a higher power level. That's why God made linear amplifiers.

To be perfectly clear, the FCC mandates that we "use the minimum power to effect and maintain communications." This doesn't mean that we need to run 1500 watts to talk to a friend across town! Conversely, it also means that if we are in QSO with someone and they are having a problem copying our 5-watt signals, then it would be perfectly justifiable to increase power a few decibels to reduce the difficult copy at the receiving end. Granted, you won't get to claim that particular QSO for a QRP award, but why make the other station suffer for your self-centered idea of ham radio. Flexibility is the key, here. Sure, it's noble to run 5 watts or less, but the real-world fact remains: there are going to be those times that you will need higher power than the QRP "legal limit." Deal with it.

SKILLS

One of the very first facts that a new QRPer has to face is, with a 13 dB power disparity between 5 and 100 watts, special skills are vital for success. Learning to listen is paramount. Listen to the DX pileup. Learn when the ebb and flow occurs, and get your call in where it will do the most good, instead of trying to go head-to-head with the wolfpack. Listen to how the other operator is working stations and time your call so he can more easily pick your QRP signals out. Listening skills are developed after spending hours at the radio.

Sending ultra-clean CW is a must. Don't try to send faster than you can copy. If you have a problem visualizing the characters as you send them, write them down first, so you don't make unnecessary mistakes. I am not a good CW op. Because I have a problem visualizing the words I want to send and invariably displace/reverse characters when trying to send from memory, I always write down specific things prior to going on the air, like my QTH, weather info, rig and antenna info, etc. This saves me the embarrassment of making mistakes and causing confusion to the other station. (This is one of the major reasons I like to send CW via a keyboard. I can type at 60 to 70 wpm with few mistakes, so it makes for cleaner CW.)

IN CONCLUSION

Ours is a great hobby and there is room enough for everyone. We show the world how to communicate, *and* we do it with only a few watts, much to the chagrin of the high-power crowd.

Appendix A

QRP Calling Frequencies

Band (m)	*CW (MHz)*	*SSB (MHz)*
160	1.810	1.910 1.843 (Europe)
80	3.560 3.579 (Colorburst Xtls) 3.710 (Novice)	3.985 3.690 (Europe)
40	7.040 7.035 (QRP-L) 7.030/7.060 (Europe) 7.110 (Novice) 7.112 (NorCal Xtls)	7.285 7.090 (Europe)
30	10.106 10.116 (QRP-L)	
20	14.060	14.285
17	18.096	18.130
15	21.060 21.110 (Novice)	21.385 21.285 (Europe)
12	24.906	24.950
10	28.060 28.110 (Novice)	28.885 28.385 (Novice/Tech+) 28.360 (Europe)
6	50.080-50.100 50.110 (Europe)	50.125 50.110 (Europe)
2	144.060	144.285 144.585 (FM)

My thanks to Bob Hightower, KI7MN, and the AZ ScQRPions for the use of this list from their Web site.

B Appendix

COMMERCIAL QRP RADIO EQUIPMENT& KIT MANUFACTURERS

Elecraft, PO Box 69, Aptos, CA 95001-0069 TEL: 831-662-8345 (**www.elecraft.com**)

ICOM America, Inc., 2380 116th Ave NE, Bellevue, WA 98004 TEL: 425-454-8155 (**www.icomamerica.com**)

Kanga US, 3521 Spring Lake Dr., Findlay, OH 45840 TEL: 419-423-4604 (**www.bright.net/~kanga/kanga**)

MFJ Enterprises, Inc., 300 Industrial Parkway Rd., Starkville, MS 39759 TEL: 800-647-1800 (**www.mfjenterprises.com**)

Oak Hills Research, 2460 South Moline Way, Aurora, CO 80014 TEL: 303-752-3382 (**www.ohr.com**)

SGC, Inc. 13737 SE 26th St., PO Box 3526, Bellevue, WA 98009 TEL: 800-259-7331, 425-746-6310 FAX: 425-746-6384 (**www.sgcworld.com**).

Small Wonder Labs (Dave Benson, K1SWL): 32 Mountain Road, Colchester, CT 06415 (**www.smallwonderlabs.com**)

Ten-Tec, 1185 Dolly Parton Parkway, Sevierville, TN 37862 TEL: 800-833-7373 (**www.tentec.com**)

Wilderness Radio(Bob “QRP Bob” Dyer): P.O. Box 3422, Joplin, MO 64803-3422 TEL: 417-782-1397 (**www.fix.net/~jparker/wild.html**)

Yaesu (Vertex Standard), 10900 Walker St, Cypress, CA 90630 (**www.vxstdusa.com**)

COMMERCIAL & KIT QRP ACCESSORIES MANUFACTURERS

Bencher, Inc: 831 N. Central Ave, Wood Dale, IL 60191
TEL: 630-238-1183 (**www.bencher.com**)

American Morse Equipment (Doug Hauff, KE6RIE):
(**www.americanmorse.com**)

Idiom Press: P.O. Box 1025, Geyserville, CA 95441,
(**www.idiompress.com**).

Jameco Electronics: 1355 Shoreway Road, Belmont, CA 94002
TEL: 800-831-4242 (**www.jameco.com**)

LDG Electronics, Inc.: 1445 Parran Rd., St. Leonard, MD 20685
TEL: 877-890-3003 (**www.ldgelectronics.com**).

Vibroplex Co, Inc.: 11 Midtown Park E., Mobile, AL 36606
TEL: 800-840-8873 (**www.vibroplex.com**)

W4RT Electronics: 3077-K Leeman Ferry Road, Huntsville,
AL 35801 (**www.w4rt.com**)

West Mountain Radio: 18 Sheehan Ave, Norwalk, CT 06854
TEL: 203-853-8080 (**www.westmountainradio.com**)

COMMERCIAL ANTENNA MANUFACTURERS/ ACCESSORIES & CABLE

Arrow Antennas, (Allen Lowe, N0IMW):
(**www.arrowantennas.com**)

Alpha Delta Communications, INC: P.O. Box 620, Manchester,
KY 40962 TEL: 888-302-8777 (**www.alphadelta.com**).

Davis RF, PO Box 730, Carlisle, MA 01741 (**www.davisrf.com**)

PAC-12 Portable QRP Antenna (James Bennett, KA5DVS):
ka5dvs@pacificantenna.com (**www.pacificantenna.com**)

The Ventenna Company, PO Box 445, Rocklin, CA 95677
TEL: 888-624-7069 (**www.ventenna.com**)

RF Connection: 213 North Frederick Ave, Gaithersburg, MD 20877
(**www.thefrc.com**)

Radio Works: Box 6159, Portsmouth, VA 23703
TEL: 800-280-8327 (**www.radioworks.com**)

Quicksilver Radio Products (John Bee, N1GNV): 30 Tremont Street, Meriden, CT 06450 (**www.qsradio.com**)

MISCELLANEOUS

EZNEC, Roy Lewallen, W7EL, P.O. Box 6658, Beaverton, OR 97007 TEL: 503-646-2885 (**eznec.com**).

EQF Software, 547 Sautter Drive, Crescent, PA 15046 TEL: 724-457-2584 (**www.eqf-software.com**).

The HW-8 Handbook, Mike Bryce, WB8VGE: **prosolar@sssnet.com**

The Heathkit Shop (**www.theheathkitshop.com**) (Everything you need or wanted to know about the Heathkit series of ham radio kits)

Glowbugs Internet Reflector (**glowbugs@piobaire.mines.uidaho.edu**): To subscribe send an email to: **majordomo@piobaire.mines.uidaho.edu** with the following command in the message body: subscribe glowbugs your email here

AMSAT-NA: 850 Sligo Avenue, Suite 600, Silver Spring, MD 20910 TEL: 888-322-6728 (**www.amsat.org**)

About The American Radio Relay League

The seed for Amateur Radio was planted in the 1890s, when Guglielmo Marconi began his experiments in wireless telegraphy. Soon he was joined by dozens, then hundreds, of others who were enthusiastic about sending and receiving messages through the air—some with a commercial interest, but others solely out of a love for this new communications medium. The United States government began licensing Amateur Radio operators in 1912.

By 1914, there were thousands of Amateur Radio operators—hams—in the United States. Hiram Percy Maxim, a leading Hartford, Connecticut inventor and industrialist, saw the need for an organization to band together this fledgling group of radio experimenters. In May 1914 he founded the American Radio Relay League (ARRL) to meet that need.

Today ARRL, with approximately 170,000 members, is the largest organization of radio amateurs in the United States. The ARRL is a not-for-profit organization that:

- promotes interest in Amateur Radio communications and experimentation
- represents US radio amateurs in legislative matters, and
- maintains fraternalism and a high standard of conduct among Amateur Radio operators.

At ARRL headquarters in the Hartford suburb of Newington, the staff helps serve the needs of members. ARRL is also International Secretariat for the International Amateur Radio Union, which is made up of similar societies in 150 countries around the world.

ARRL publishes the monthly journal *QST*, as well as newsletters and many publications covering all aspects of Amateur Radio. Its headquarters station, W1AW, transmits bulletins of interest to radio amateurs and Morse code practice sessions. The ARRL also coordinates an extensive field organization, which includes volunteers who provide technical information and other support services for radio amateurs as well as communications for public-service activities. In addition, ARRL represents US amateurs with the Federal Communications Commission and other government agencies in the US and abroad.

Membership in ARRL means much more than receiving *QST* each month. In addition to the services already described, ARRL offers membership services on a personal level, such as the ARRL Volunteer Examiner Coordinator Program and a QSL bureau.

Full ARRL membership (available only to licensed radio amateurs) gives you a voice in how the affairs of the organization are governed. ARRL policy is set by a Board of Directors (one from each of 15 Divisions). Each year, one-third of the ARRL Board of Directors stands for election by the full members they represent. The day-to-day operation of ARRL HQ is managed by an Executive Vice President and his staff.

No matter what aspect of Amateur Radio attracts you, ARRL membership is relevant and important. There would be no Amateur Radio as we know it today were it not for the ARRL. We would be happy to welcome you as a member! (An Amateur Radio license is not required for Associate Membership.) For more information about ARRL and answers to any questions you may have about Amateur Radio, write or call:

ARRL—The national association for Amateur Radio
225 Main Street
Newington CT 06111-1494
Voice: 860-594-0200
Fax: 860-594-0259
E-mail: **hq@arrl.org**
Internet: **www.arrl.org/**

Prospective new amateurs call (toll-free):
800-32-NEW HAM (800-326-3942)
You can also contact us via e-mail at **newham@arrl.org**
or check out *ARRLWeb* at **www.arrl.org/**

FEEDBACK

Please use this form to give us your comments on this book and what you'd like to see in future editions, or e-mail us at **pubsfdbk@arrl.org** (publications feedback). If you use e-mail, please include your name, call, e-mail address and the book title, edition and printing in the body of your message. Also, please indicate whether or not you are an ARRL member.

Where did you purchase this book?

☐ From ARRL directly ☐ From an ARRL dealer

Is there a dealer who carries ARRL publications within:

☐ 5 miles ☐ 15 miles ☐ 30 miles of your location? ☐ Not sure

License class:

☐ Novice ☐ Technician ☐ Technician Plus ☐ General ☐ Advanced ☐ Amateur Extra

---

Name ____________________ ARRL member? ☐ Yes ☐ No

Call Sign ____________________

Daytime Phone ( ) ____________________ Age ____________________

Address ____________________ e-mail ____________________

City, State/Province, ZIP/Postal Code ____________________

If licensed, how long? ____________________

Other hobbies ____________________

____

Occupation ____________________

For ARRL use only	**LPC**
Edition	2 3 4 5 6 7 8 9 10 11 12
Printing	1 2 3 4 5 6 7 8 9 10 11 12

From ______________________________

____

____

Please affix postage. Post Office will not deliver without postage.

EDITOR, LOW POWER COMMUNICATION
AMERICAN RADIO RELAY LEAGUE
225 MAIN STREET
NEWINGTON CT 06111-1494

please fold and tape

THEY'RE A WEIRD MOB